DRINK?

The New Science of Alcohol + Your Health

PROFESSOR DAVID NUTT

yellow
kite

First published in Great Britain in 2020 by Yellow Kite
An imprint of Hodder & Stoughton
An Hachette UK company

1

Co-author: Brigid Moss

Cover design and Illustration by Emily Langford © Hodder and Stoughton
Figures © 2019 Cali Mackrill at Malik & Mack

A CIP catalogue record for this title is available from the British Library

Hardback ISBN 9781529393231
eBook ISBN 9781529394726

Typeset in Sabon MT by Hewer Text UK Ltd, Edinburgh
Printed and bound in Great Britain by Clays Ltd, Elcograf S.p.A.

Hodder & Stoughton policy is to use papers that are natural, renewable
and recyclable products and made from wood grown in sustainable
forests. The logging and manufacturing processes are expected to
conform to the environmental regulations of the country of origin.

Yellow Kite
Hodder & Stoughton Ltd
Carmelite House
50 Victoria Embankment
London EC4Y 0DZ

www.yellowkitebooks.co.uk

CONTENTS

PREFACE

Most people know me as the UK government chief drugs advisor who was sacked in 2009 for saying government drugs policy wasn't evidence-based. But I'm also a doctor and all my professional life I, like all doctors, have been confronted with the challenge of alcohol in our patients and in our colleagues.

The impact of alcohol on health is profound. Up to half of all people in beds in orthopaedic wards are there because of alcohol-related injury and at weekends the A&E departments of our hospitals are filled with people who are drunk. Alcohol and medicine are inextricably entwined in the same way as money and banking.

The real reason for my sacking, I believe, was that I had the temerity to say on primetime radio that alcohol was the most harmful drug in the UK. At that time alcohol wasn't even allowed to be considered a drug by the UK Advisory Council on the Misuse of Drugs – the ACMD – despite every scientist in the country knowing that it most certainly was a drug.

The evidence on which I based my statement was the most sophisticated and detailed analysis of drug harms ever conducted. Since then, similar studies have been carried out in Europe and Australia, each coming to the same conclusion: that alcohol is the most harmful drug in their societies, too.

The main reason why alcohol scores at the top of the harm scale is that so many of us like to drink it. Typically, in first world western countries alcohol is consumed by over 80 per cent of all adults. Of that 80 per cent only about one-fifth get into problems with it. But there are so many of these people that their alcohol problems have a massive impact on the rest of us, especially their families and friends. Alcohol is linked to a lot of violence both within and outside of the family, road traffic accidents, loss of work and many illnesses. Policing drunkenness in the UK costs over £6 billion a year and the health costs are over £3 billion.

Yet most of us continue to drink, and most don't get into serious difficulties with alcohol. This tells us that there are different biological and social factors that impact our relationship with alcohol. I believe that understanding these can help each of us, and our governments, make more rational and health-promoting decisions about how we all deal with alcohol. This book attempts to do this in a language every drinker can easily understand.

At a personal level, alcohol has been something I have been studying over my 40 years of medical research. For two years, in the late 1980s, I ran the inpatient research ward at the National Institute on Alcohol Abuse and Alcoholism (NIAAA) in the National Institutes of Health (NIH) in Maryland, USA. Since then, back in the UK, I have continued to study the brain basis of alcohol enjoyment and problems, along with treating patients with alcohol problems.

Yet also, with one of my daughters, I own a wine bar in Ealing, London. My life encapsulates the good and bad of alcohol and pulls together my knowledge, experiences and ideas in a way that I hope will explain why such a simple molecule as alcohol can give so much pleasure and pain at the same time.

Professor David Nutt is Professor of Neuropsychopharmacology at Imperial College London, Chair of DrugScience (drugscience. org.uk) and a former chair of the Advisory Council on the Misuse of Drugs.

INTRODUCTION

It was my first day at Cambridge, and somehow all of us first-year medical students in the same college found each other. Nine of us went to the pub around the corner.

We settled in for the evening. It was a typical 1960s city pub: small sticky tables, even stickier carpet, with a smoky atmosphere that had given a yellow veneer to the walls and ceiling. I wasn't a big drinker, but I had, I think, three pints of the local Greene King beer. Most of the others drank maybe four or five, which seemed a heck of a lot to me. At 10.30 p.m., the pub closed. Someone said they had a few bottles of wine back in their room, so we went back to the common room and hung around drinking them.

The mood was pretty loud and raucous, a bunch of green eighteen-year-olds having a laugh. But then, one of the guys started to cry. Not quietly but very loudly, wailing and weeping. It was an enormous outpouring of negative emotion so intense that I thought he was going to kill himself. I asked a friend who'd been at school with him, should we call an ambulance? But he said, 'No, don't worry, he always does this. He won't even remember it in the morning.'

He didn't remember it. But I couldn't help wondering: how could alcohol turn someone who's a high-achieving school success, on the surface confident and likeable, into a gibbering mess? I began to think, what does this tell us about alcohol? Looking back, that was the beginning of my fascination with drugs, including alcohol, and how they affect the brain, the person and society at large.

As a doctor, I've seen first-hand all the types of havoc that alcohol causes, from liver disease and cancer to drink-driving accidents and violence. What is clear is that you don't have to be a homeless alcoholic on the street to be harming your body and/or brain with booze. You can be a high achiever, like my university friend.

In fact, he wasn't the only one of that small group who had problems with alcohol. The school friend who reassured me he would be OK was also a heavy drinker. Very sadly, he had problems with alcohol his whole life, and died of liver failure in his forties, his promising career and indeed life ruined by alcohol. True, theirs are extreme stories. But they're not particularly unusual ones.

Now, 50 years on, I want to share what I've learned about alcohol. I'm not here to preach, and I do not judge you, whether you're drinking too much or you're worried that you are; I've written this book to help you make informed and smart drinking choices. Even while I've been writing, it's made me rethink, again, how and what I drink myself. I'm not a big drinker and I know all the risks, but even so I can see how easy it is to slip into the habit of overindulging.

My position has always been: alcohol is a drug and one that shouldn't be taken lightly. This may sound odd in our culture, where having a drink and indeed getting drunk is broadly acceptable. But it's a position most doctors, scientists and addiction experts take too, given the enormous amount of damage alcohol does to people's health and to society.[1]

The World Health Organization's *Global Status Report on Alcohol and Health 2018* said that in 2016 more than 3 million people died as a result of harmful use of alcohol. A recent editorial in the *British Medical Journal* (*BMJ*) sums up how: 'Like tobacco, alcohol kills some users slowly through the diseases it causes. Unlike tobacco, alcohol also kills quickly, through injury and poisoning.'[2] Alcohol causes more than 5 per cent of the global disease burden. It's estimated that alcohol contributes to the death of around 30,000 people a year in England, Scotland and Wales.[3,4,5,6,7] The big killers are cardiovascular disease, which leads to heart attacks and stroke. Then there is liver disease and various cancers, as well as accidents, especially on the road; and deaths from suicide.

I'm not trying to scare you, I just want you know the facts, so you can make your own decisions. It's true that since around 2010, as in most of Europe, our drinking habits have been going in the right direction, which is down. But that's from a very heavy baseline, as we in the UK drink nearly twice what we drank in the 1960s. Now the proportion of us drinking more than 14 units a week is still one in five of us – nearly a third of men and one in six women.[8]

In the UK, we are particularly good at drinking. According to the Global Drug Survey, Britons get drunk an average of 51.1 times a year – that is, once a week. And one in ten adults are drunk on five or more days a week.

Two-thirds of people who drink over 14 units say they'd find cutting down on drinking harder to do than other lifestyle changes like exercising more, reducing smoking or improving their diet. If you have picked up this book, I'd imagine you might be one of them. Perhaps, in the past, you've felt relieved when you've seen a newspaper article about 'moderate' drinking, particularly red wine, being good for our health. But the truth is, as you'll see in Chapter Two, no level of alcohol consumption is safe.

You can dress alcohol up as an Aperol spritz or a glass of Pol Roger, or dress it down as a two-litre bottle of extra-strength white cider, but either way, it's a molecule that taps into our brain chemistry to produce a whole range of effects. Including, in some people, addiction. Until you appreciate that, you run the risk of underestimating its power.

The shift in the way we consume alcohol in the last 50 years, the length of my career, is a remarkable example of how marketing has altered our perception. Alcohol used to be a special purchase, and you had to go to a special shop – an off-licence or the pub – during limited hours to buy it. Since successive governments from the 1970s onwards liberalised our access to buying alcohol, as well as massively lengthening the hours alcohol could be sold, we now chuck it in our supermarket basket or trolley as part of our weekly shop.

It's not until you think about it that you realise how alcohol is entrenched in every part of our lives. We drink for social bonding. We drink together to clinch business deals and come to agreements. We drink to celebrate the birth of a child, to commiserate with each other when someone dies.

When one of my nephews turned 18, I went to buy him a birthday card. I counted 23 eighteenth cards before I found one that didn't focus on alcohol. What kind of message is that for a young person? Luckily, it seems as if that age group – or some of them – are drinking less. Because it's absurd that coming of age should be all about alcohol. Still, those cards show that reaching the age of majority when you are allowed to drink is seen as one of the great transition points in a person's life.

Alcohol has glamour and history, and our art and culture are steeped in it. It's thought to be nearly as old as human society. There's a theory that the roots of agriculture weren't in the search for food but in the cultivation of crops to make alcohol.

The Romans had their Bacchanalia, the Greeks their symposia, both fuelled by wine. We have our hen and stag parties, our wine societies and beer clubs, office Christmas parties and dinner parties.

In the thirteenth and fourteenth centuries, the Catholic Church was the main source of alcohol as monks held a lot of the cultural knowledge of wine- and beer-making. The oldest continuously produced beer goes back nearly a thousand years and is from a monastery in Germany.[9] But now, one company produces a third of the beer in the world.[10]

Drinking – and drinking heavily – is in our cultural references, from the paintings of Toulouse-Lautrec to James Bond. Toulouse-Lautrec's favourite drink was reportedly absinthe, the Parisian drink of the moment, mixed with brandy. And if I drank as many Martinis as 007, I'd be shaken too. He should be called the man with the golden liver!

Sure, not all cultural references are positive. In the novel *L'Assommoir* by Émile Zola, a woman desperately tries to hold her family together while her husband becomes an alcoholic, but she ends up dying in penury as one herself. In the critically acclaimed film *Judy*, Renée Zellweger brings to life the sadness of the final alcoholic years of Judy Garland.

The only difference between alcohol and any other drugs, I've always argued, is that alcohol is legal. That position has got me into trouble – neither the government nor the drinks industry want alcohol to be treated as a drug, for obvious reasons – but I am a scientist and when you look at how it affects the neurotransmitters of the brain (see Chapter One), it's clear it's pretty close in its brain effects to quite a few other drugs; although it's more promiscuous than any other. No other drug both enhances GABA, serotonin and dopamine and at the same time blocks glutamate and noradrenaline. Alcohol truly produces a cocktail of neurotransmitter effects.

In three papers I co-wrote between 2010 and 2019,[11] we showed alcohol is the most harmful drug in the UK, Europe, Australia and so probably in all the western world. We did a decision analysis to rank legal and illegal substances, including heroin, cocaine, crack and crystal meth, as well as alcohol and tobacco. Alcohol scored the highest not because of the harm it does to the individual but mainly because of the vast harm it does to others – passive drinking, if you like (more on this in Chapter Eleven).

In my first ever proper job as a psychiatry trainee, working at the A&E department at Guy's Hospital in south-east London, one of my very first patients arrived in a terrible state, borderline psychotic. She'd come straight from her wedding reception, where, in a drunken brawl, her brothers had beaten up her new husband. Sadly, that is a story repeated in various forms in A&E departments across the UK on Friday and Saturday nights.

INTRODUCTION

It is true the huge amount of harm from alcohol is largely driven by the number of people who drink. If half of adults took crystal meth, for example, society would be in a much worse state. But on a personal level, there are very many harms of alcohol and these are determined by how much you drink. They may not be obvious, like the link to breast cancer that has only recently become clear, or the 10 to 20 years that drinking takes to damage your liver.

You may be hoping none of this will apply to you. Most of us love alcohol's short-term effects – the fun, the sociability, the relaxation – and hope the long-term health damage won't happen to us.

But if you are drinking above (or even below) 14 units (112 grams) a week, you need to know the risk to your health. I hope this book will make this clear. The drinks industry knows alcohol is a toxic substance. If it were discovered today, it would be illegal as a foodstuff. The safe limit of alcohol, if you applied food standards criteria, would be one glass of wine a year.[11] Would you take a new drug if you were told it would increase your risk of cancer, dementia, heart disease, or that it shortened your life? You wouldn't touch it. But alcohol has a special place in our culture.

What I would like to help you do is make you think about your drinking. To know what the risks are if you go above the recommended limits. To know that even if you can't stick to the limits all the time, you should not just give up and ignore them completely. To learn how to monitor your drinking. Perhaps, sometimes, to choose not to drink.

What, ideally, I don't want you to do any more, is to throw alcohol down your throat without thinking. When you've read this book, you may decide to join the growing number of people who decide not to drink. Or you may decide you are happy drinking exactly as much as you are now, but you will have made that decision in possession of all the facts.

Because what drinking alcohol needs to be, above all, is a conscious act. You should treat it as more special than eating, more like – I would argue – it used to be treated in the past. Make drinking a positive, active pleasure rather than a reflex and habit, or something you've always done, or self-medication for stress or anxiety.

My view is, try to reduce how much you are drinking as much as you can towards the recommended limit while maximising your fun and pleasure. And have at least two days off a week. I hope I'll give you ideas for how to do that.

I do not blame you if you drink too much. Not only does society introduce us to drinking at an early age and tell us it's the most fun we can have, but as you'll find out (see Chapter Seven), it's very easy to become psychologically and/or physically addicted to alcohol.

I hope this book will show you how to drink for living rather than to live – or die – for drinking. I hope it will help you start to take back your agency over how much you drink, to be able to decide how much is right for you – then drink exactly that. Cheers to that!

1

HOW DRINKING AFFECTS YOUR BODY AND BRAIN

FROM FIRST SIP TO THE MORNING AFTER: ALCOHOL'S JOURNEY

LET'S TAKE A short tour through your brain and body on alcohol. Your first drink starts with a single mouthful, to bastardise a saying by Mao Tse Tsung.

In fact, you would be surprised how difficult it is to get people to drink alcohol for the first time. Pure alcohol tastes repellent, and nobody – except perhaps an alcoholic – would drink it. Even in the various forms in which we buy it, it's often bitter and tart, if not downright unpleasant, and needs a mixer. Some of the common flavours that alcohol comes in – wine-flavoured, hops-flavoured – are quite strange and so also have to be acquired.

One of the most effective ways to make alcohol more appealing to people is to add sweeteners. This holds true for rats, too. When scientists use rats to study the effects of alcohol, it's standard to sweeten it to ensure the rats drink it; a rat alcopop, if you like. In fact, I'd say the main reason supersweet (and often brightly coloured) alcopops are on sale is to appeal to entry-level consumers who don't yet have the taste for alcohol, aka teenagers.

You may also be surprised to hear that we quickly acquire that taste. Very soon after the first mouthfuls, our brain learns that a few minutes after the difficult taste come good feelings of being warm, relaxed, maybe a little more sociable. And once those good feelings become associated with the taste, the smell and the setting, routine and/or ritual, we then start to like the taste. I've often heard people say things along these lines: 'I love the taste of 1984 Château Latour/Whispering Angel/Pol Roger.' But I say to them: 'If you gave that to your

IS MORE EXPENSIVE BOOZE BETTER FOR YOUR BODY?

You can't buy your way out of harming your body. There may be an argument that red wine and beer contain some health-protective factors (see Chapter Two for more), but they don't contain enough to make up for the harms of ethanol, alcohol's most toxic chemical. If you get less of a hangover when you upgrade your drinks, it may be that you are savouring it more and so drinking less; which is no bad thing. But sorry, the short answer is no.

child, they would spit it out. You have acquired the love of that taste. And what has given you that love of the taste is the effect of the alcohol. And, of course, the knowledge that it's really expensive.'

This learning effect holds true in rats too: once a rat has been made to drink enough alcohol, it will continue to drink it, even if it's not flavoured.

Another sensation many people learn to like is the hot feeling you get as alcohol goes down your throat, which will be familiar if you drink spirits and especially if you've ever done shots. It can become appealing because your brain knows you're about to get the pleasurable hit from the alcohol. As spirits are a dilute form of alcohol, that hot feeling is not nearly as intense as if you were to put neat alcohol on your tongue or a cut. That would burn – and really hurt.

These two elements – the flavours and the mouth-feel as well as the look of your favourite tipple and the place and time you usually drink – prompt your brain to get ready to experience the effects of alcohol.

As the liquid flows into your stomach, it begins to be absorbed through the walls of the stomach, then via the small intestine. The alcohol goes via the bloodstream into the liver, where it starts to be broken down, the main by-product being acetaldehyde. Then this alcohol and acetaldehyde mixture travels through the bloodstream and into the heart and also crosses the blood–brain barrier and enters your brain.

Alcohol is such a small molecule that absorption happens pretty quickly; within five or ten minutes of your first mouthful, you will start to feel the first effects in your body. You may begin to feel warmer, a little flushed, as alcohol allows the blood vessels of your skin to expand, which is called vasodilation.

INSIDE YOUR BRAIN

When the alcohol mixture hits the brain, it starts to do the things that make us like it. I find it helpful to describe the brain as working like an electrochemical machine. This machine is made up of a web of around 200 billion neurones. All our thoughts and processes are mediated by messages between the neurones. Your machine's outputs – being awake, asleep, storing memories, swallowing and so on – are the results of these millions of messages, flying around the brain web.

Each message travels along the neurone via electricity, but the connection that bridges the gap between neurones – called the synapse – is chemical. The chemicals that bridge those gaps are called neurotransmitters.

What's important to know for the story of how alcohol works is that it – like other drugs – works at the level of this chemical connection.

There are around eighty different types of neurotransmitters (chemical messengers). And there are even more types of the receptors that they slot into. Each receptor is triggered by a different neurotransmitter.

The two neurotransmitters that are the most common and the most powerful, because they are effectively the on–off switch of the brain, are gamma-aminobutyric acid (GABA) and glutamate. In essence these two neurotransmitters are the core of the brain. They do all the basic work such as sleeping, laying down memories and thinking.

Glutamate turns on the brain and GABA turns it off. When glutamate is released and goes across the synapse it turns on the next neurone, which makes the brain more active. GABA does the opposite.

The two work hand in hand, like yin and yang. You need GABA to control glutamate, because if you have too much glutamate your system goes into overload; this leads to getting very anxious, or even to a seizure, and potentially to brain damage.

So your brain has evolved so that every time glutamate is released, GABA is released too. It's a beautifully balanced system.

There's another type of neurotransmitter, called a neuromodulator because it modifies the brain's response rather than affecting it directly. So, if you have a car accident, the memory of it will be laid down by glutamate – for example that you were driving at 60 miles an hour, that you were on the A34 – but the emotion will be encoded by a neuromodulator, noradrenaline. This adds extra information to the memory but it's not the core component.

Neuromodulators work like a backup to GABA and glutamate because most outputs of the brain use more than one neurotransmitter.

To explain with an example, we think there are at least four neurotransmitters that keep us awake: acetylcholine, orexin, histamine, noradrenaline – and at least four that put us to sleep: GABA, adenosine, serotonin, endocannabinoids. This means that if one fails, the others can back it up. So someone with narcolepsy, for example, has a deficiency in one neuromodulator, orexin. They're not good at staying awake because the backups aren't always powerful enough.

The names of some neuromodulators will be very familiar to you. As well as noradrenaline, serotonin and dopamine are involved in many processes, particularly emotional ones. Then there are the endorphins, which are actually a different type of molecule called a peptide, little pieces of protein.

All the various drugs and mind-altering substances that human beings take work on different combinations of neurotransmitter and neuromodulator systems. Alcohol is one of the most promiscuous of drugs, in that it affects a lot of different types of receptors and hence the majority, if not all, of the neurones. Which is why it can give us so many different kinds of effects and experiences.

YOUR BRAIN ON ALCOHOL

As you can see, the brain is an incredibly evolved and finely balanced machine. And then you add in alcohol and that balance dissolves like the sugar cube in a hot cup of tea – or rather, a hot toddy.

The first thing alcohol does is to turn on the calming GABA system, so you start to feel relaxed. Valium works on the same system. This is why we drink. And it's especially why a lot of us drink at parties.

Most of us have some level of social anxiety, and alcohol removes our fear and inhibitions. Alcohol's calming effect is also the reason why, as soon as the seatbelt sign is turned off on an aeroplane, the drinks trolley is wheeled down the aisle. It's because many people are anxious about flying.

However, if you turn the GABA system on too much, it can switch off parts of the brain you don't want switched off, for example your judgement or even your consciousness. And if you drink a great deal of alcohol and the GABA system is maximally potentiated, it turns off your brain, in the same way as an anaesthetic, so you stop breathing. That's one way you can die from alcohol. This is what we call alcohol poisoning.

Going back to your judgement, do you ever intend to drink a couple of drinks but then lose control and end up bingeing? Why is that? One of the main reasons is that the part of the brain that tells you to stay in control – the frontal cortex – is the first part that's switched off by alcohol.

In fact, the parts of the brain that are about retaining judgement and control are very sensitive to alcohol. You may find you lose judgement over other things too: your attractiveness, your ability to dance or to chat someone up.

Drink more, and as you go over the drink-driving limit – which is 80 milligrams of alcohol per 100 millilitres of blood (80mg%) – you begin to get the double whammy: as well as stimulating GABA, the alcohol starts to block your glutamate receptors. And remember, glutamate is the neurotransmitter that keeps you awake. As your level gets higher, you're starting to become properly drunk. If you reach the point of 150mg%, you'll also start to lose the capacity to lay down memory. This is a blackout.

As your blood alcohol concentration rises, it affects new and different neuromodulators too. And each one will have its own specific influences in different parts of the brain.

There are three that are the most important. Firstly, a rising blood alcohol level increases the effects of serotonin, a mood enhancer that also makes you more empathetic. Its pro-serotonin effect is also what makes other people seem more attractive – so-called 'beer goggles'. In that way, it has a similar effect to MDMA or ecstasy.

It's the stimulation of a different serotonin receptor, one in the nerves of the stomach, that makes you sick.

We talk about throwing up from alcohol pretty casually but, in fact, vomiting is crucial as it stops you from dying. This survival mechanism is one of the reasons alcohol has survived in our culture for thousands of years. Vomiting gets rid of enough alcohol so you stay alive.

Secondly, drinking releases dopamine, which is involved in drive, motivation and energy. This is a factor in alcohol's stimulant effect, which makes you feel exhilarated, more active, gives you feelings of energy and enthusiasm. Dopamine makes you louder – this is an effect people get from cocaine, too.

Dopamine is one of the transmitters that also lays down behavioural patterns, so it's important in addiction. Dopamine release may be the reason people get locked into habits that start off being fun, or at least not damaging, but then become so, for example hair-twisting and nail-biting. Dopamine is also why you get into stupid arguments about irrelevant things when you're drunk, but can't stop yourself.

Thirdly, the high you get from alcohol comes from endorphins. These are your body's natural opioids, the brain's natural pain-reducing system, also the source of the runner's high. This reward system gives you a chilled sort of pleasure and, in some people, may also be a key factor in their addiction. Several studies have shown that when the effects of endorphins are blocked with the anti-drinking medication nalmefene, some addicts are able to stop drinking.[1] Using sophisticated brain imaging, we have been able to see that this effect is related to the interactions between endorphins and dopamine in the brain.

It's this cornucopia of effects that gives alcohol such a wide appeal. And as we are all different, it will slot into your personal brain chemistry in a different way to that of your friend. Perhaps you drink to reduce anxiety? Or after work as a valve to release stress? Or to get motivated to go out? Or for Dutch courage to go fighting? A large part of alcohol's appeal is that, for many people, it fills in the gaps in your personality, making you the person you want to be.

THE MIND-BENDERS IN YOUR BOOZE

What you drink is as full of psychoactive substances – chemicals that change your brain chemistry – as any other drug. By far the most abundant alcohol in the bottles on off-licence shelves is ethanol (chemical formula C_2H_5OH). During fermentation, the glucose in the raw ingredients breaks down mostly into ethanol, but it will always make other types of alcohol too, which differ in amounts and chemical structure in different drinks. Even what is sold to us as the purest alcohol, for example vodka, contains a cocktail of various alcohols. The only pure alcohol is ethanol that's been synthetically produced.

So, for example, when whisky has been analysed, it has been found to contain roughly 400 different alcohols. This is because the longer a whisky is stored – or 'aged' – the more some of the alcohols will join together to form more complex alcohols called congeners. These exist in all alcoholic drinks (although whisky contains the most) and are thought to work similarly in the brain to simple ethyl-alcohol (ethanol), but may perhaps be even more intoxicating. Each whisky will have its own combination of alcohols and congeners and it's this mixture that creates all the various nuances of its 'nose' and so its flavour.

Interestingly, under the Psychoactive Substances Act 2016, every recreational psychoactive substance was banned except ethanol, caffeine, nicotine and tobacco. At the time of the Act, I did say that this was absurd. In effect it makes all alcohol illegal, because there is not a single alcohol you can buy that doesn't contain congeners. Beer has around 150, wine has 200. We don't drink pure ethanol.

As well as their own blend of hundreds of different alcohols, wine, beer and cider also contain aromatic plant compounds called terpenes, which come from grapes, hops or apples. You may have heard of terpenes as they give cannabis its distinctive smell, too. It used to be assumed that terpenes contributed purely to flavour but it's now thought possible – though it isn't yet well studied – that they are also psychoactive.

SO YOU KEEP DRINKING . . .

Keep drinking and, by this point, you are likely to be slurring. Your memory is shot due to the glutamate effect, so you're repeating yourself, repeating yourself.

You may also be finding a lot of things funny, including the things you're repeating, which may be down to your serotonin and GABA systems. And alcohol is dampening down the centres of the brain that control coordination, hence the slang 'legless'.

With your dopamine up and your self-control down, you may be beginning to get argumentative. You may also do something stupid – smoking when you've given up, drink-driving, shopping-trolley racing – because your judgement is impaired.

There was a really sad case of a 16-year-old girl, Natalie Dursley, who was picked up by an ambulance after collapsing at a nightclub. She was so confused, she opened the doors of the vehicle on the M5, fell out and died. This is a perfect example of the complete loss of insight that can happen when you are drunk. At root, you've seriously disrupted the efficiency of your brain – an effect I like to compare to having a virus in your computer.

Keep on drinking, and you're moving towards anaesthesia, a system shutdown. In fact, medical anaesthetics target the GABA and glutamate systems too. An anaesthetic you'd have for a minor op would switch on GABA to put you to sleep. And for a major one, it'd switch off glutamate, which keeps you awake and alive. That is why you need to be ventilated, as you can no longer breathe on your own.

Before modern anaesthetics existed, if a sailor had to have an injured limb removed, he'd be made legless first. There's some good reason for this: alcohol does dampen down the pain – as well as the memory.

It's the fact that alcohol affects both GABA and glutamate – the double whammy – that makes it dangerous. GABA and glutamate not only regulate being awake but also being alive. This is why people do die of alcohol poisoning. If you drink enough, you can stop breathing. Below is an at-a-glance guide to how drunkenness progresses.

WHAT HAPPENS AS YOU DRINK MORE? THE STAGES OF DRUNKENNESS

As your blood alcohol content rises, it affects the way you feel and behave, as well as the different neurotransmitters in your body.

blood alcohol concentration / you will feel / neurotransmitter(s)

- 20mg% / relaxed, slightly altered mood, slightly warmer / GABA? – endorphins
- 50mg% / less inhibition, louder speech, more gesturing, reduced coordination and eye focus / GABA – serotonin
- 80mg% / drink-driving limit. Loss of coordination, balance, speech, hearing and reaction times / GABA – endorphins
- 100mg% / slurred speech. Reduced reaction times and physical control / GABA – dopamine
- 150mg% / euphoria, coordination reduced so much that you may fall over. Walking and talking become hard. Possibly vomiting / Serotonin – dopamine – GABA
- 250mg% / confusion, stupor, disorientation. Loss of pain response. Nausea and vomiting may begin. Hard to stand and walk without help. Blackouts begin / glutamate block + GABA
- 300–400mg% / Possibility of falling unconscious / glutamate block + GABA
- 400mg% / Stupor/ glutamate block + GABA
- 500mg% / Coma / glutamate block + GABA
- 600mg% / Breathing stops and death / glutamate block +GABA[2]

CONTEXT IS EVERYTHING

Different people do have very different experiences under the influence, which comes from the setting and expectations, more so than from any difference in the alcohol itself. The buzz from a warm can of gin and tonic on a packed commuter train will feel completely different from the same measure in a Martini in a smart hotel bar, for example. And preloading before clubbing is a world away from a glass of sherry with your parents at Christmas. People might drink at bedtime to relax or they might (though it's best not to) drink before a business meeting for confidence.

Buckfast Tonic Wine is interesting as a social experiment (albeit inadvertent) and a great illustration of how alcohol's effects interact with personal intention. Originally made by the monks of Buckfast Abbey in Devon as a pick-me-up, 'Buckie' is a strong, fortified wine – 15 per cent alcohol – with lots of added caffeine. One bottle contains the equivalent of nearly five double espressos.

It was first sold as a tonic for people who needed energy, as the caffeine overcomes the sedative effects of the alcohol. You might know that Buckie has become the cult drink of football followers in Scotland (where it is also known as 'wreck the hoose juice'); these fans down a bottle (or maybe more) before going out for a fight. In a Scottish survey of young offenders, 43 per cent who'd drunk before their offence said their drink of choice was Buckfast.[3]

It's the same wine as when it was marketed as a tonic, but is now used for a completely different purpose. In 2014, the Scottish government debated whether Buckie should no longer be sold in glass bottles, as there were reports of them being used as weapons, according to the *Scotsman* newspaper.[4] Although this law was not passed and Buckie still comes in a wine bottle, it is now – perhaps to reduce the risk of broken bottle violence – available in cans as well.

Of course, violence and alcohol are well-established bedfellows. It's one reason most sporting venues and festivals no longer use glasses (the other is that drunk people are so clumsy).

TIMING IS ALL

The time of day you drink is also a factor. It makes sense that the sleepier you are, the more likely it is that alcohol will put you to sleep. That's the reason people mix stimulants and alcohol as in Buckie; stimulants to keep you awake, sedatives to take away the anxiety. In fact, 'uppers and downers' is the most popular combination in the history of drug-taking.

Another example of this is cocaine and alcohol. Back in the 1890s, when cocaine was legal, a wine called Mariani from Italy contained both. And it was endorsed by the Pope, no less. And you may have heard the rumours about David Cameron, Boris Johnson and other Bullingdon Club members allegedly taking cocaine and drinking alcohol at parties.[5] The reason why people might do this is in order to be able to drink more and for longer. Interestingly, when in the 1990s the Icelandic government passed a law to allow 24-hour drinking, there was subsequently an increase in amphetamine use.[6]

One of the issues with cocaine and alcohol is that they work together in the body to produce a new chemical, called coca-ethylene (CE). CE is a longer-acting form of cocaine that hangs around in the body for hours rather than minutes. And this makes it more toxic to the heart. That's why there's such a strong association between taking cocaine, drinking and heart attacks.

The upper–downer effect accounts for the popularity of Red Bull and vodka, too, otherwise known as a Vodka Rush; it's a mix that's been shown in animal studies to have brain-altering effects.[7] You might consider a Vodka Rush terribly unsophisticated and prefer an espresso Martini, but in reality, what you're doing to your brain is not so different.

Alcohol with tobacco is probably the most common combination of all. A lot of people find that once they've got into the habit of having a cigarette with a drink, it's hard to have one without the other. This may be down to the fact that smoking accentuates the impact of alcohol on dopamine. It may also be down to the fact that alcohol disinhibits you, so after a drink you lose the will not to smoke.

There is no doubt, either, that getting drunk faster – and so raising your blood alcohol levels faster – will mean you overcome your inhibitions more quickly. Drink fast, and you may find your insight has gone and you are overwhelmed by alcohol before you even realise you are drunk. The strength of the alcohol you're drinking also matters in this respect; the faster you are able to get drunk, the more quickly bad things start to happen.

This is why the trend for drinks becoming stronger in the past fifty years or so is not a positive one. In the 1960s and 70s, when I was a student, most lagers and ales in pubs were 3 to

4 per cent. Then along came Stella at 5 per cent. Now pub beers are routinely 4 to 5 per cent. Wine's strength has also generally gone up, from 11 or 12 per cent to 13 or 14.

Some drinks have been shown to make your blood alcohol levels rise faster for other reasons too: champagne more than wine, for example. In fact, most people get drunk faster on fizzy drinks.

THE HIDDEN DANGERS OF TOLERANCE

The toxicity of alcohol will vary according to your genetics but also your past drinking experience. This is your tolerance, your body and brain's ability to prepare for and withstand alcohol. Tolerance builds up very fast because your brain quickly learns to expect alcohol. At the start of a holiday you might feel drunk on, say, a glass or two of wine. By the end of the week, it might take a bottle to feel the same way.

A high tolerance always has its dangers. I once went to stay with a friend, and when we sat down to watch the rugby he offered me a drink. When I said yes, he poured me a full wine glass of Scotch, which I sipped. He poured himself the same measure – and, while I sipped mine – kept finishing it and pouring himself more. By the end of the match, he'd drunk a full bottle of whisky (minus my one glass!). His tolerance was sky high.

The sad story of Amy Winehouse illustrates another danger of tolerance: the consequences of losing it. Coming out of six weeks of being in treatment, she went back to drinking the amount she had done before rehab. But because she'd lost her tolerance, this amount of alcohol had become enough to poison her.

HOW DRINKING AFFECTS YOUR BODY AND BRAIN

If you think about it, tolerance is really dependency – or at least the beginnings of it. If you look back and realise your drinking has been steadily increasing, take a look at Chapter Seven, where I explain more about how tolerance works.

WHY DO I GET DRUNK BUT MY FRIEND DOESN'T?

There could be a few reasons – and there are likely to be more than one. Not least, there's tolerance: your friend might be used to drinking a lot more than you, so their brain will be prepared to combat the effects of alcohol. If your friend is a different sex to you, that will make a difference. If a man and a woman drink the same, the woman's blood alcohol level will be higher. It's not only because women tend to be smaller but also that, proportionally, women's bodies have a higher percentage of fat and a lower percentage of water, and alcohol is diluted in the body's water content.

How much your friend has eaten and what they've eaten will make a difference too, as food delays alcohol uptake from the stomach. This is the reason why some people swear by drinking a pint of milk before going out on a heavy night. And it's not just perceived wisdom. True, the milk won't have a huge effect, but the fluid will fill them up and hydrate them so they're less thirsty. The fat, as with other fat-containing foods, delays stomach-emptying. As only 20 per cent of alcohol is absorbed in the stomach, this slows alcohol uptake into the blood.[8,9]

Then there are genetic factors. Your friend's genetics may mean they metabolise alcohol faster than you. Some people have trouble metabolising alcohol, in particular the stage after the body turns it into acetaldehyde. This inability is most common in people of Japanese, Chinese and Korean descent.[10] The result is, they get the bad effects fast, in particular flushing and feeling very drunk quickly.

Some people, particularly the sons of male alcoholics, seem to inherit brains that are sub-sensitive to alcohol. This means that they can tolerate more than their peers from the very first time they drink. It's a kind of pre-tolerance that seems good to start with but over time can lead them to drink more and so be at a greater risk of alcohol damage.

THE AFTERMATH ...

WHY YOU EAT MORE WHEN DRUNK

Do you often end the night going for a kebab or a burger or your drunk food of choice? It's not well understood why we get so hungry when we've been drinking.

One reason might be that alcohol changes the production of sugar in the liver so you become hypoglycaemic; in other words, you get low blood sugar so you feel hungry. Or it may be the case that eating, like smoking, is a way of boosting your reward-pleasure experience. Or if you are dieting or restricting food in some way, it could be that alcohol's inhibition-reducing effect on GABA reduces your resolve not to eat. There's probably a similar reason why drunk people are more likely to have sex too.

WHY YOU WAKE UP TIRED AND WIRED

As it's a sedative, alcohol puts you into very deep sleep. This explains why, after they fall asleep drunk, people do odd things, such as get up and pee in the wardrobe. They are in such a deep sleep that their conscious brain stays asleep even while the part of their brain that's telling them to pee is awake. And as night terrors – intense nightmare-like experiences where you can still move – happen in this deep sleep state too, if you're prone to them, alcohol can provoke them (more on this in Chapter Six).

The sedative effect works for around four hours. At this point, you may wake up with a headache and perhaps with lights flashing in your eyes, feeling alert and unhappy and find it hard to get back to sleep. If you don't wake up, you're likely to sleep badly or fitfully.

This is happening because your brain knows being intoxicated is potentially dangerous – so it doesn't want to be intoxicated. As soon as you started to drink the previous night, your brain started to change. Because alcohol blocks glutamate receptors, your brain up-regulated the glutamate system to compensate for this, by increasing the quantity of glutamate receptors.

Then, as your blood alcohol level goes down during the night, you're left with too many receptors and so too much glutamate activity. And that is why you are too awake and alert, and why the world seems too bright, too noisy, too much.

WELCOME TO YOUR HANGOVER

Good morning! Or likely not. People joke about hangovers – 'What's the best thing for a hangover? Drinking heavily the night before.' There's even a multimillion-dollar film franchise based on the confusion, sickness and regret we experience the morning after. But what is happening is more serious than that. It's withdrawal. Yes, the symptoms are less extensive, unpleasant and life-threatening than an alcoholic will go through. But however you dress it up, your brain and body are withdrawing from alcohol.

The definition of an alcohol hangover is the experience of various unpleasant physiological and psychological effects that follow the medium-to-high consumption of alcohol. Typically, it comes on around ten hours after your blood alcohol peaks – but this varies according to sex, weight and your genetic disposition too. Some people don't get them at all: between 3 per cent[11] and 23 per cent[12] of the population are reported to be hangover-resistant. Not surprisingly, these people may be more likely to become heavy drinkers as they don't experience the deterrent effect of hangovers.

As you might have experienced, hangovers can last from a few hours to over 24. And as alcohol affects so many systems in your body and brain, hangover is equally complex, hence the myriad of possible ways to suffer. There are over 47 possible symptoms (see table on page 31), and you're likely to get your own personal selection of them; but they range from sleep disturbance and dehydration to anxiety, tension and negative emotional state, to impairment of attention, memory and psychomotor skills. Symptoms will depend on how much you've drunk but also on what you've drunk. Cask-aged and matured alcohols have a lot of congeners (which we looked at earlier), and it is thought that these may create worse hangovers.

The upshot? Even though you're no longer drunk, you probably won't be much use at work and you almost definitely shouldn't be driving. The US Centers for Disease Control and Prevention estimated that alcohol hangovers cost the US economy approximately $249 billion in 2010.[13]

WHAT'S HAPPENING IN YOUR BRAIN AND BODY

There are gaps in the research about what is going on in your body during a hangover. What we do know is that the hangover state is a multifactorial event caused by a variety of biochemical and neurochemical changes as well as your personal genetic make-up.

1) **YOU'VE POISONED YOURSELF** One of the ways alcohol is metabolised is by the enzyme alcohol dehydrogenase (ADH). As this enzyme breaks down ethanol, it forms acetaldehyde, a poison and carcinogen. This is relatively quickly turned into acetate, then finally into carbon dioxide and water. However, some people have genetic variants of the relevant enzymes that make the breakdown faster or slower. What's known from studying people who are slow to break down acetaldehyde is how unpleasant it makes you feel: flushed, nauseous, rapid heartbeat.[14] So some of the symptoms of hangover may come from acetaldehyde hanging around.

Consuming high quantities of congeners is thought to make hangovers worse, too. There are high concentrations in red wine and distilled spirits, for example brandy, and low ones in clear spirits such as vodka. Congeners include acetaldehyde, acetones, histamines and methanol. In fact methanol, a product of sugar fermentation, is thought to be a major contributor to the symptoms of hangover. Alcohol dehydrogenase, ADH, will metabolise methanol at a slower rate than ethanol to form formaldehyde (which is used to preserve bodies) and formic acid (found in the stings of bees and ants), both of which are highly toxic.

2) YOUR NEUROTRANSMITTERS HAVE GONE HAYWIRE I've already explained about the changes in glutamate and GABA, your main neurotransmitter systems. The degree of the imbalance between them – i.e. too much glutamate and too little GABA – has in rodents been shown to correlate with the intensity of withdrawal.[15]

3) YOU ARE INFLAMED The inflammatory response happens when your body is damaged, as part of the immune system's response. Despite the fact that it's a natural response, it can be very destructive. Chronic inflammation is now believed to be a significant factor in many long-term health conditions, from diabetes to cancer and, as we shall see in the next chapter, liver cirrhosis.

Alcohol turns on this process because it damages the blood vessels and your gut, so the body then turns on itself. The inflammatory response is unpleasant – symptoms can include nausea, vomiting, headache, confusion and tremor, as well as clinical depression, which induces mood changes and cognitive impairment, and learning and memory deficits. This is why an anti-inflammatory medicine – for example ibuprofen – can help the symptoms of hangover.

4) MITOCHONDRIAL DYSFUNCTION Alcohol also damages mitochondrial DNA, particularly in the liver. Mitochondria are the energy-producing machines in every cell and are susceptible to damage from free radicals produced by alcohol via acetaldehyde. Brain cells are reliant on mitochondria for their energy supply and even slight damage to the mitochondria can lead to toxicity in a number of brain regions.

THE 47 SYMPTOMS OF HANGOVER

In a survey of 1,410 Dutch students, these were the symptoms they named. They're listed in order of how commonly they occurred. So 95.5 per cent of those surveyed said they felt fatigued, while only 1.8 per cent said they had suicidal thoughts.[16]

Fatigue (being tired)	95.5
Thirst	89.1
Drowsiness	88.3
Sleepiness	87.7
Headache	87.2
Dry mouth	83.0
Nausea	81.4
Weakness	79.9
Reduced alertness	78.5
Concentration problems	77.6
Apathy (lack of interest/concern)	74.0
Increased reaction time	74.0
Reduced appetite	61.9
Clumsiness	51.4
Agitation	49.5
Vertigo	48.0
Memory problems	47.6
Gastrointestinal complaints	46.7

DRINK?

Dizziness	46.0
Stomach pain	44.7
Tremor	38.9
Balance problems	38.6
Restlessness	36.8
Shivering	34.4
Sweating	33.9
Disorientation	33.8
Audio-sensitivity	33.3
Photosensitivity	33.1
Blunted affect (displays less emotion)	29.9
Muscle pain	29.4
Loss of taste	28.0
Regret	27.1
Confusion	25.8
Guilt	25.2
Gastritis	23.4
Impulsivity	22.7
Hot/cold flashes	21.4
Vomiting	20.8
Heart pounding	19.4
Depression	18.9
Palpitations	17.0
Tinnitus	16.8
Nystagmus (involuntary repetitive eye movements)	16.1
Anger	10.1
Respiratory problems	9.7
Anxiety	7.4
Suicidal thoughts	1.8

WHERE DO HANGOVER SYMPTOMS COME FROM?

CAN'T BEAR LIGHT AND SOUND: This is down to the glutamate rebound effect.

HEADACHE: May be for the same reason as above, plus there's also the fact that acetaldehyde dilates the blood vessels in your head. New research also suggests alcohol may cause your immune system to attack your body. It seems alcohol can turn on your immune cells; they are supposed to protect you from bugs and toxins, and they do so by releasing chemicals that cause irritation in blood vessels and nerves, which can lead to pain including headaches.

There's also an element of dehydration due to the diuretic effect of alcohol (meaning that it makes you pee more). You could try water-loading before going to bed, but of course it disrupts your sleep if you have to get up to pee.

STOMACH SYMPTOMS: Alcohol damages the stomach and intestine lining and so can both give you diarrhoea and leave you feeling nauseous. There may be a little inflammation in your pancreas too – pancreatitis, inflammation of the pancreas, is commonly associated with heavy drinking.

HOW TO PREVENT A HANGOVER

1. DRINK LESS. You know this already.

2. PACE YOURSELF. It takes roughly one hour for you to process one unit (10ml/8g) of alcohol. Drink slowly and you will have a chance to process the alcohol, which means your blood alcohol peak won't be as high, which has been shown to give less of a hangover.

3. DRINK WATER WITH YOUR ALCOHOLIC DRINKS. Other ways to dilute are to add more ice, add soda to wine or lemonade to beer, or have more mixers with spirits. NB: flat is better as fizzy drinks may make you absorb alcohol faster.

4. DRINK SMALLER MEASURES. Shrinking your glass means less booze over the course of an evening.[17]

5. DRINK CLEAR SPIRITS. They contain fewer congeners, which are thought to make hangovers worse. One study showed that vodka produced a lesser hangover than whisky, for example (although both the vodka and whisky drinkers' sleep and next-day concentration were bad).[18]

HANGOVER CURES: WHAT'S THE EVIDENCE?

You can't 'cure' a hangover. All you can do at this point is treat the symptoms.

1 Anti-inflammatory. Take ibuprofen when you go to bed and/or when you wake up, ideally with food as it can be harsh on the stomach lining.

2 Beta blocker. (NB: these need to be prescribed and they are NOT licensed to treat hangovers!). This will slow down the pounding feeling in your head by reducing your heart rate. Whether they do more than that, I'm not sure.

3 Hair of the dog. This is the very worst thing you can do. You are only delaying your hangover by putting alcohol back into your brain. No stars!

4 Hydration. A pint of water before bedtime and one when you get up. Electrolyte/sports drinks also

restore salts; there's no research but the theory makes sense. I wouldn't recommend getting hooked up to an IV hydration bag unless you're in hospital, though.

5 Food. You'll likely wake up with low blood sugar, which is why eating carbs makes you feel better. Eggs contain a lot of the nutrients you need at this point.

6 Caffeine. Any source of caffeine – if you can stomach it – will help you feel more awake. Coca-Cola (flat or not) provides three things that may help: carbs, fluids, caffeine.

7 Probiotics. Probiotics are known to reduce leaky gut and a very small study on heavy drinkers showed the subjects had reduced levels of inflammation in the body.[19] Promising.

8 Herbs/supplements. Most of these have no evidence. It may be worth taking a B vitamin complex. Some companies – such as Agrip – are now developing all-in-one combination treatments to speed up alcohol metabolism (see agripdrink.com). Expensive but promising.

9 Move. In theory, any kind of exercise will help you speed up your metabolism and so help shift your hangover. However, you are unlikely to feel like doing this.

WHY YOU SHOULD BE HAPPY IF YOU GET A HANGOVER

Sensible drinking is about restraint, which may become easier when you know that alcohol impairs you the next day. In the 1980s, I used to run the National Institutes of Health inpatient research ward on alcohol in the USA. What was interesting was that, when I interviewed the alcoholics we treated and researched there, almost none of them had ever experienced a hangover. So you could theorise that they'd never had a deterrent to drinking too much.

However, in an online study of Dutch students aged 18 to 30 years who'd recently had a hangover, only a small proportion – 13.4 per cent – said that if a hangover cure existed, they would increase the amount they drank. Most – 71.6 per cent – said they wouldn't increase their alcohol consumption.[20] But of course this study asked the respondents what they thought they'd do, not what they'd actually do.

So, all in all, though getting a hangover may mean you're less likely to become an alcoholic, I hope all the other information in this chapter will help you to see how potentially dangerous too much alcohol can be on its journey through the body and brain.

2

THE HEALTH HARMS
OF ALCOHOL

WHAT'S A SAFE LEVEL?

IS THERE ANY such thing as a safe level of alcohol? This is a question I get asked all the time.

I'd like to throw that question back to you: what do you, as an individual who is responsible for your own health, accept in terms of risk?

After all, if I told you that if you drank a gin and tonic today, you'd die tomorrow, you wouldn't drink it. You wouldn't take a poison pill, would you?

But what if I told you this: if you drink half a bottle of wine a day for 40 years, it will take a year off your life?

The reality of alcohol is this: drinking raises your risk of health harms. But the impact on your health of your chosen level of drinking is not certain. What is known is on a medical and population level; we know there is a strong relationship between alcohol consumption and dying younger. Heavier drinkers die sooner, from a range of disorders. It's proven that alcohol affects more than 200 of the different diseases identified in the *International Statistical Classification of Diseases and Related Health Problems (ICD)*. Alcohol use is one of the top five causes of disease and disability in almost all countries in Europe. In the UK, alcohol is now the leading cause of death in men between the ages of 16 and 54 years, accounting for over 20 per cent of the total. More than three-quarters of liver cirrhosis deaths, 7 per cent of cancer deaths and 25 per cent of injury deaths in adults under 65 years of age in Europe in 2004 were estimated to be due to alcohol.[1] According to the government, alcohol is the third leading risk factor for death and disability after smoking and obesity. That's a pretty scary list, right?

That said, on an individual level the picture is in no way clear. Some people who drink will die from a disease, for example cirrhosis, that is directly linked to their consumption of alcohol. Some people will die earlier because alcohol has contributed to a condition, for example from having a stroke. Some people who reach a century swear that it's their daily tipple that explains their advanced years.

When it comes to all drugs, including alcohol – and in fact all risky activities – less is always safer. Looking at the statistics, if you want to maximise your life, the rule would be not to drink a drop.

The same goes for if you want to maximise your health: don't drink at all – because there are no health benefits to it. But if you want the sociability benefits alcohol brings, it's a different story.

In that case, you need to decide what risks you want to accept, balanced out with the pleasure you gain. The risks are determined by how old you are, your sex, your genetics; but, most of all, how much you drink and how often you drink. What you can do is to work out the dose that gives you the best fun but with a reasonable amount of risk – in your opinion. It's not a one-size-fits-all rule. And it really is your call.

By the time you've read this book, the idea is that you'll be able to say: I have made the decision to drink this amount. I have decided that the benefits more than offset my fear of the harms.

I want you to be able to work out how little you need in order to get the effects you do want.

But first, you need to know what all the risks of drinking are and all the potential harms. That's what this chapter is about: giving you the facts on how alcohol affects our physical health.

One of the issues is that I don't think the real risks from drinking have been made clear to most of us. We are told by the government that 14 units a week is a 'low-risk' amount, for both men and women. This limit isn't arbitrary; it was established by a group of experts based on all the available evidence, so it's very solid. But did you know this: if you stick to these levels, your risk of dying due to an alcohol-related condition is around or a little under 1 per cent. The guidelines explain that this level of risk is comparable to other regular risks, such as driving. One per cent is the level of risk the experts think is acceptable.

But the question for you is, is that the level of risk you're prepared to accept?

In the past, we've been told that drinking – specifically red wine, but also other types of alcohol – is good for cardiovascular health. This is, of course, an extremely appealing message. But do you perhaps use this as an excuse to drink too much? More recently, the increased risk of cancer from drinking has become clear, and it has also emerged that, on balance, no level of drinking is actually beneficial to health.

One study concluded that even a couple of drinks more than four days a week raises your risk of premature death by 20 per cent.[2]

Perhaps you drink more than 14 units a week and want to know what the risks are for you? Perhaps, when you have all the information, you might decide to cut down or stop drinking?

The main thing is to know what you are doing – and that is what I aim to help you to do. It's up to you: after all, as I said, there is no level of alcohol consumption that is without risk. However, when you drink low amounts, the risks of health harms are low too.

Another factor that makes the risk to health from drinking so difficult to assess is because there are a lot of conditions it can lead to, each of them with its own risk, from liver cirrhosis to various cancers and cardiovascular disease.

Contrast this with smoking, which over 50 years of data has been shown to be the main cause of lung cancer and puts you at high risk of cardiovascular disease. So while smoking

THE HEALTH HARMS OF ALCOHOL

leads to a high risk of two conditions, drinking over the safe limits doesn't give such a high risk of any one condition – but it is involved in many more conditions.

WHAT ARE THE RISKS OF DRINKING?

We all know smoking and being overweight is bad for us, and if we smoke or overeat, we know we shouldn't.

But drinking alcohol also has the same kind of long-term or chronic risk, which means it kills you slowly. The eminent statistician David Spiegelhalter uses the concept of the microlife to explain how fast it does this.

One microlife is 30 minutes off your life expectancy. If you are a 30-year-old man, you will lose one microlife by smoking two cigarettes, or each day of being 5kg overweight, or for every seven units of alcohol. Seven units might be: two pints of strong 6 per cent beer or three pints of 4 per cent beer, three-and-a-half double gin and tonics (25ml pub measures), or three medium glasses (175ml) of 13 per cent wine. But it gets worse if you drink more than that because of the exponential rise in alcohol harms – see the table overleaf.

ARE YOU SHORTENING YOUR LIFE?

This is an estimate of the impact that drinking at different levels will have on your life expectancy.

It's a calculation based on averages, which means there's no guarantee it will apply exactly to you. But it's useful to know, as at least it gives some proportion to the amount of risk you are taking.

You'll see that the more you drink, the – much – worse it gets because of the exponential nature of the health harms of alcohol. That is, the harm increases much faster than the amount you're drinking.

Here, the impact of alcohol and other risky things is explained in terms of losing microlives.[3] This is a period of half an hour, and is used because 1,000,000 half-hours = 57 years, which roughly corresponds to an adult lifetime of exposure. The following gives an indication of the amount of life lost by drinking each week at different levels over a period of 40 years. I've included some estimates for smoking and being overweight, too.

Intake	life shortened in years
14 units (current UK limit for men and women)	0.4 years
21 units (previous UK limit for men)	0.8 years
35 units (half a bottle of wine a day)	2 years
70 units (1 bottle of wine a day)	7 years
140 units (2 bottles of wine a day)	21 years
20 cigarettes a day	8 years
Being 5 BMI units a day over a body mass of 22.5 units, i.e. BMI of 27	2.4 years off
Being 5kg overweight	0.9 years

In addition, the risks from alcohol get bigger the more you drink. In the figure below, you can see that for both men and women the

THE HEALTH HARMS OF ALCOHOL

increase is not linear but exponential, i.e. the risks get proportionally bigger the more you drink. The arrows shows that the impact of reducing alcohol consumption from 100 to 50g a day actually decreases the risk of death from 16 to 2. In other words, halving consumption decreases the risk by eight times.

Fig 1. Reducing high-end consumption has major impact on harms

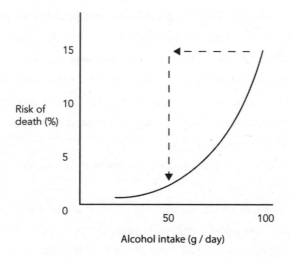

Source: https://journals.sagepub.com/doi/abs/10.1177/0269881113512038

Another issue is that the link between drinking and disease isn't pointed out to us by government or doctors nearly as clearly as smoking: there aren't pictures of diseased livers or breast cancer tumours on the label of a bottle of Pinot Noir from Sainsbury's, for example.

And a complicating factor is that while any level of smoking – from just one cigarette – increases your risk of lung cancer and heart disease, the same isn't true of every condition.

For smoking, the graph that shows risk looks like a straight line, rising proportionally the more you do it. But the relationship between alcohol and risk varies according to each condition. For cardiovascular issues, problems start at two or more units (one pint of beer) a day, and accelerate when you drink at a higher level. This relationship isn't a straight line but rather looks like a curve, especially when it comes to heavy drinking (see table).

The good news is, the shape of the curve means that, if you drink a high amount, cutting down will reduce your risk of harm much more than if you are reducing from a low level (see table). So, for example, a reduction of four units a day when you're drinking a lot – say from ten units a day down to six – will have a much bigger impact on health than dropping from five down to one.

What this means in practice is, if you do drink a lot, it really is worth reducing it.

The fact that cardiovascular health harms don't start at zero has given us a feeling that you can drink 'safely'. However, in the 21st century, researchers have begun to uncover more about the relationship between drinking and, particularly, cancer, where the harms appear to start at a much lower level. For this and possibly (debatably) for brain health, the more you smoke or drink, starting from any level of intake, the more likely you are to get the health condition.

Furthermore, historically most of us have grown up with the idea of a 'safe limit' of drinking as laid down by the government. Those past recommendations were based on what was known of the harms then – mainly liver damage and cardiovascular effects. But this language has left us thinking that if 14 units a week is 'safe', how can two or three or even five a week more be much worse?

We know more now. That's why the safe limits, which are set by the UK Chief Medical Officer, are now called the 'low-risk drinking guidelines'.

As of January 2016, these guidelines have been set at 14 units a week for both men and women, spread over a few days, with several drink-free days a week.[4] In practice, this looks like two pints of beer at 4 per cent or two 175ml glasses of 13 per cent wine a day, spread out over three days a week, with days off in between. At lower levels men and women have a similar risk, but men's risk increases faster as consumption goes up – which is one of the reasons why men's limits were lowered to 14 units, the same as women's.[5]

These guidelines were developed in consultation with experts, who looked at all the existing evidence. And in fact, they do now state clearly that no level of drinking is safe when it comes to cancer.

Interestingly, despite governments worldwide having access to the same science, guidelines vary. The standard drink size or unit – 8g of alcohol in the UK – varies in different countries from 8g to 20g. And low-risk drinking guidelines vary hugely too, from 10–42g per day for women and 10–56g per day for men; while weekly guidelines vary from 98–140g for women and 150–280g for men.[6]

What factors will affect your risk of health harms?
- How often you drink
- How much you drink

- Your general health. The poorer your health for other reasons, the more alcohol may affect you.
- Your age. The older you are, the more vulnerable you become to health harms.
- Your sex. Women and men have different health risks.
- Your genetic inheritance.
- The number of years you've been drinking. Most often, health issues due to alcohol only begin to show up ten to twenty years after drinking began.
- The age you began drinking. The earlier you began drinking, the more you may have affected your health.
- Whether your family has a history of alcoholism; if it does, you are more likely to become dependent.

WHAT IS A BINGE?

This means getting drunk, as opposed to just having a fun time with alcohol. It implies being intoxicated to a point of being out of control – which in some people means that they lose control of their drinking, then drink massively more than they intended or is good for them.

The NHS defines it as: 'drinking lots of alcohol in a short space of time or drinking to get drunk'. And the official classification of a binge is six units in a single session for women or eight for men.

Not only does bingeing lead to more problems when drunk – for example doing things that may have a negative impact on your work, your relationships and even your life, from things like peeing in public to very serious conse-quences such as molesting other people – but it can have significantly more health impacts than drinking the same amount over a few days. For example, a study from Sussex University found bingeing damages the brain more than the same amount of alcohol taken in a more spread-out fashion.[7]

A binge can kill you, even if you don't feel particularly drunk. Women are more likely than men to die from a binge. Being on average smaller means that for any given alcohol intake their blood alcohol levels will be higher. That was the sad story of Paula Bishop who, in 2017 over the course of one day on holiday in Fuerteventura, drank a few small beers, two glasses of wine and four Irish coffees; it's thought around 15 units. She had a blood alcohol level of 4 grams per litre – five times the drink-driving limit in the UK. But, her husband said, she hadn't appeared particularly drunk.[8] (NB: there's another lesson here – measures can be a lot bigger in other countries).

Bingeing on alcohol can also put you at risk of being harmed or assaulted. That's because when you are this drunk, you become unaware of what is going on around you.

It's also likely that bingeing makes the drinker more vulnerable to becoming addicted because it produces more extreme effects in the brain. That means the brain has to produce more adaptive changes to compensate. I think these changes may one day be proven to be the start of a slippery slope to dependence on alcohol.

Some people, especially younger people, intentionally binge. They drink to get 'hammered' or 'wasted', to lose control, to forget their everyday life. But there are other people, who binge despite not wanting to. They go out for 'just a couple' of drinks and then lose control. These people can be helped by treatment with nalmefene, a medicine that blocks the endorphin rush from alcohol that contributes to loss of control and hence more drinking.

THE MAJOR WAYS ALCOHOL AFFECTS YOUR LENGTH OF LIFE

LIVER

When people think about alcohol and health, this is what they think of first. Because everyone knows drinking too much damages your liver. In fact, it's not the most prevalent cause of harm from alcohol – that accolade goes to cardiovascular conditions. But it's certainly a very important one and, sadly, one that's becoming increasingly common.[9]

One feature of liver cirrhosis is that the person's abdomen swells up and collects water. A standard piece of medical student education was to examine these patients with one hand on one side of the abdomen and then tap the other side with the free hand. This tap sent a wave through the fluid, which could be felt by the other hand (we call this a thrill). When I was learning medicine such patients were quite rare overall but relatively common in France, so our name for alcoholic cirrhosis was 'the French disease'. Since then, the French government has introduced policies that have significantly reduced cirrhosis rates, whereas our policies have led to it rising (more on this in Chapter Nine). So now it's the English – or sometimes Scottish – disease.

Alcohol is a poison, and the liver is in charge of breaking down poisons, as well as other drugs and chemicals. That's why alcohol, like other poisons, can destroy your liver. The problem is, most people don't know they have liver damage until it's very advanced – and possibly too late.

1) FATTY LIVER The first stage of liver disease is called hepatic steatosis, which means fatty liver. The liver gets fatty because it makes fat from the calories in the alcohol, and stores it in its cells. Your liver may also be fatty if you are overweight, have insulin resistance or are diabetic – this type of non-alcoholic fatty liver disease is very common.

One issue is that most people don't feel any symptoms during this early stage of liver damage. You might feel a little abdominal discomfort over your liver – which is in the upper right of your abdomen – perhaps a little heartburn or nausea.

If a GP thinks you may have fatty liver, they will do blood tests. You may also be offered a Fibroscan, a specialist ultrasound that can measure scarring and fatty change in your liver.

It's been shown that a lot of middle-aged 'average' drinkers have fatty liver. According to the charity Drinkaware, you are at risk if you drink more than eight units a day if you are a man and five if you are a woman – for just two or three weeks, which is only as long as a long holiday! NHS advice says if you stop drinking for two weeks, you can reverse fatty liver disease (after this, you need to carry on drinking within the guidelines, of course).[9] But the time it takes to recover will depend on the severity of your condition, among other factors, especially being overweight.

2) ALCOHOLIC HEPATITIS Most of the alcohol you drink is broken down in the liver, into energy plus acetaldehyde, which at one carbon atom longer than formaldehyde (used to preserve Damien Hirst's sharks and cows) is even more of a toxic pickling agent than alcohol. This poison also kick-starts the process of inflammation.

If you go on a significant binge – maybe half to a whole bottle of spirits or ten units in the course of an evening – you can get acute alcoholic hepatitis. Your liver will become inflamed by the vast amount of alcohol you have taken. Occasionally people die of a very fatty liver that becomes inflamed.

It's more common to develop acute alcoholic hepatitis over three to four days of drinking. We now see this condition in young binge-drinkers; in fact, there are wards full of these kids. There is a particular danger point for dying the night you become a legal drinker – 18 in the UK, 21 in the US – because people press drinks on the young person, thinking they are being nice. For the parents of these children, it becomes the worst day of all.

Binge-drinking is particularly dangerous because when the liver dies, you tend to die. The only way to recover from it is by having a liver transplant. However bingeing is not the reason behind most cases of alcoholic hepatitis, it is diagnosed much more often in long-term heavy drinkers.

There's also another type of hepatitis that occurs in drinkers: autoimmune hepatitis. When your liver starts to get damaged by alcohol, and inflamed, it becomes more vulnerable to attack by the immune system.

Finally, if you have some other form of hepatitis that's not caused by alcohol – for example hepatitis A, B or C – then drinking alcohol massively increases the rate of dying from this.

3) CIRRHOSIS This is the most serious form of liver damage. The liver has quite a lot of regenerative capacity – which is why you can recover from fatty liver – but there comes a point of no return. And cirrhosis is it.

In the same way you might pickle fresh fruit in alcohol to make the German pudding *rumtopf*, alcohol pickles your liver. Cirrhosis happens when, over decades, the mixture of alcohol and acetaldehyde permeates the tissues of the liver and begins to freeze all the metabolic activity. The tissues can't even degenerate because the enzymes that usually digest the dead cells, the ones that are also responsible for the breakdown of a dead body, are killed too. In this state, the liver gradually solidifies.

As you become more tolerant to alcohol and the amount you drink increases, more and more of the liver dies, so that instead of being a soft, squidgy organ that's full of blood, it becomes dry and hard and woody.

It loses its capacity to make the things you need it to make, such as enzymes and some hormones. In fact, when your liver starts to pack up, your alcohol tolerance disappears because it can no longer metabolise the alcohol. That's when you start to get drunk on small amounts of alcohol. If it becomes clear this is what is happening, you must stop drinking immediately. You are facing terminal liver failure.

As the liver becomes more and more blocked by this fibrous tissue, it becomes harder for blood to flow through it. As most of the blood from your gut goes through the liver, it has to go somewhere, so it starts going through little vessels in the walls of the gut and the gullet (oesophagus), which start to carry far more blood than they were designed for. These veins get bigger and bulgier. The bulges are called varices, and are similar to the bulges found in leg veins when you have varicose veins.

Often, the first time someone finds out they have serious liver disease is when they start vomiting blood when one of these vessels has burst. In hospital, an operation or a pressure tube in the oesophagus can stop the vessels bleeding, but the more serious issue is that by that stage, the liver is usually beyond repair.

Towards the end of the cirrhosis process, you can no longer make the enzymes you need to digest your food. Another key function of the liver is to make the proteins that your body needs to make blood. When your liver isn't working, your protein levels go down and you get the fluid build-up in your abdomen that I described above. It's this metabolic disturbance that eventually stops your heart.

If you do have some liver function left, you can stay alive with cirrhosis for some years. But you are extremely vulnerable to catching a fatal infection; your body easily becomes overwhelmed as you don't have the right level of proteins in your blood.

The other thing the liver does is break down the toxins produced by the microbiome so that they don't get to the brain. That's why people at this stage get confused and delirious – a phenomenon called hepatic encephalopathy – because their brain is being poisoned by these substances. Doctors can give people antibiotics to kill the gut bugs, but this only gives temporary respite. At this point, life expectancy is very short – unless you have a transplant.

George Best, a long-term alcoholic, was a classic case of cirrhosis. He not only killed his own liver, but he was lucky enough to be given a transplant – and then killed that one, too.

You might be thinking, is it appropriate to give such a scarce and precious resource to a person who may drink again, and so squander it? I think about it like this: the vast majority of people who drink don't end up dependent. And as alcohol is legal, you can't blame the people who try it, then do get addicted. They didn't intend to become alcoholic; they are the unlucky ones. Should you penalise someone for being unlucky?

You don't have to be the cliché of an alcoholic who is drinking day and night to die of cirrhosis. A small study by the UK's leading alcohol and liver disease expert, Professor Nick Sheron, found that a third of patients with severe alcohol-induced liver damage had never even considered that they were drinking abnormally. They met the criteria for excessive drinking without ever having realised it. Only 9 per cent showed evidence of severe alcohol dependence. Most had not lost control of their drinking; rather they were classified as heavy drinkers.[10] Professor Sheron says: 'The majority of patients presenting with alcoholic liver disease appear to be heavy controlled or social drinkers, leading relatively controlled lives, perhaps not feeling that their drinking is a major health issue until they are diagnosed with end-stage liver disease, at which point the liver has been damaged to the extent that only 30 per cent will be long-term survivors.'

We really do not understand why some people get cirrhosis and others do not. We presume it's something to do with the body's immune response and genetics and is likely to be aggravated by poor diet.

Patterns of drinking seem to matter, too. According to the Million Women Study, daily drinkers have a higher risk of cirrhosis than non-daily drinkers, and women who drink mainly with meals have a lower risk than those who drink without food.[11] It also seems that women may be more susceptible to the toxic effects of alcohol on the liver.[12]

But the important message for everyone is: you don't have to be dependent in the sense that you crave alcohol or have trouble stopping drinking in order for it to kill you.

4) LIVER CANCER Any chronic inflammation can lead to cancer. So one of the other possible consequences of cirrhosis is primary liver cancer – cancer that starts in the liver. If you have cirrhosis, you have a 12 per cent lifetime risk of developing liver cancer. It is very hard to treat and is usually a terminal diagnosis.[13]

OTHER CANCERS

This is the – pretty long and scary – list of other cancers for which alcohol is known to be a significant risk factor: breast, colorectal, oesophageal, pharynx and larynx, lip and oral cavity and nasal. Some cancers are dose-dependent (see box overleaf) but some aren't.

The causal association is strong enough that alcoholic drinks have been classified as a carcinogen by the International Agency for Research on Cancer. And it's also why the World Health Organization has decided that no level of alcohol consumption is safe.

That doesn't mean there aren't other cancers where alcohol contributes. For example, World Cancer Research Fund says there is limited but 'suggestive' evidence that drinking may contribute to cancers of the lungs, pancreas and skin.[14] And we're finding out more all the time; for example, it was only in 2017 that the full extent of the association of drinking and breast cancer became clear. According to a World Cancer Research Fund report from that year, just a single drink a day increases the risk of breast cancer.[15]

There is some good news: a moderate intake of alcohol (under 3.75 units a day) appears to lower the risk of kidney cancer.

There are two ways in which alcohol increases the risk of cancer.

Firstly, there is the acute toxic effect of alcohol on the skin of your mouth and gullet, where it burns and damages the tissue. This chronic inflammation leads to cancer through changing DNA, particularly once it has led to persistent mouth ulcers that last for months. While it's not known why some people go on to develop, for example, mouth cancer, it is known that drinking is a major contributory factor (others are smoking and the human papilloma virus or HPV).

The second way is by a more indirect route, which will vary for each type of cancer. This process is thought to involve the production in cells of molecules called free radicals, which then leads to the DNA damage that promotes cancer.

DRINK AND CANCER: HOW MUCH ALCOHOL CAUSES CANCER?

There is strong evidence that:

Consuming any alcohol at all increases the risk of mouth, pharynx and larynx cancers, oesophageal cancer (squamous cell carcinoma) and breast cancer (pre- and post-menopause)

Consuming 3.75 or more units a day (30 grams or more) increases the risk of colorectal cancer

Consuming 5.5 units a day (45 grams or more) increases the risk of stomach cancer and liver cancer

Consuming up to 3.75 units a day (up to 30 grams) decreases the risk of kidney cancer

---

How much alcohol do you need to consume to increase the risk of cancer? This statement comes from AMPHORA, a recent, very detailed European Commission report on the many effects of alcohol.[16] NB: This statement also refers to all the other carcinogens in alcoholic drinks. But as they're present at levels much lower than ethanol itself, it's the ethanol that's the main carcinogen.

Ethanol is a carcinogen, a teratogen and toxic to many body organs. Using the European Food Standards Authority guidance on risky exposure for human consumption of toxic substances in food and drink products, European drinkers consume more than 600 times the exposure level for geno-toxic carcinogens, which is set at 50 milligrams (0.05g/day) alcohol per day; and more than 100 times the exposure level for non-carcinogenic toxins, which is set at 0.3 grams alcohol per day. (The average consumption of the 89 per cent of EU citizens who drink alcohol is just over 30g/day.)[17]

This means that to stay below the acceptable food standard risk for cancer you should consume no more than 50mg/day. This works out as 18g/year i.e. a maximum of two units a year!

CARDIOVASCULAR HEALTH AND BLOOD PRESSURE

There are a few ways in which alcohol affects cardiovascular health. Drinking can damage the heart, a condition called cardiomyopathy. It weakens the heart muscle so it becomes less efficient, less powerful at pumping the blood. It is really poisoning your heart, in the same way as it poisons your liver. Over many years, this can eventually lead to heart failure.

Even light to moderate drinking raises your risk of an irregular heartbeat, cardiac arrythmia, which may make you feel faint and short of breath.[18]

More importantly, alcohol is also the leading preventable cause of hypertension (high blood pressure). According to US figures, drinking contributes to at least 7 per cent of cases. Why is increased blood pressure dangerous? Because it leads to two main outcomes: heart attacks and strokes.

We don't fully understand why drinking raises blood pressure. One reason might be that during alcohol withdrawal – which, you will remember, you go through every time you drink – there's a surge in the neurotransmitter noradrenaline, and that's known to increase blood pressure.

It used to be thought that alcohol provided some protection against stroke. In the past, this has been the source of countless headlines about the benefits of red wine. Because everyone who drinks loves to read all the reasons why it's a good idea, right?

There's now controversy over whether alcohol does provide any protection. A 2018 review in *The Lancet*, one of the leading medical journals, was called – pretty definitively – 'No Level of Alcohol Consumption Improves Health'.[19] Their conclusion was that, on balance, any protection would be more than cancelled out by the negative effects.

The Lancet article has been criticised for stating this as fact, when the way the studies are set up – they are observational – means they can only say there is a correlation between a level of drinking and a disease, rather than a cause. It also didn't differentiate between what people were drinking: whether it was wine, lager, scrumpy or vodka, for example.[20]

The critics of this report aren't saying that heavy drinking isn't bad for you: that is well established. But what we don't know is what happens to the health risks between no drinks and one or two daily drinks – and one reason for this is that the difference in risk is so small.

There may well be a partial protective effect on cardiovascular health – *The Lancet* paper suggests this is true for women, at least, though previous research has pointed up this effect in men too. But there are two really important points to mention. The first is that the amount of alcohol optimal to provide the protection appears to be very low – about one unit a day. Secondly, that benefit to the heart does not outweigh all the other risks of alcohol, most notably cancer.

When it comes to alcohol, it's been shown that the risk of stroke is both dose-dependent and cumulative.[21] I don't think most people who are being treated for hypertension are even asked whether they're drinking too much, I suspect because doctors feel uncomfortable asking this – perhaps because many doctors drink too much themselves?

There is some good news. It seems the effect on blood pressure is reversible if you stop drinking. In one study,[22] 13 out of 18 heavy drinkers saw their blood pressure return to normal levels after one month of abstinence. However, if you're a heavy drinker, you'll need to cut down slowly rather than cut it out immediately, or you may get a rebound effect of very high blood pressure from the noradrenaline surge that comes with withdrawal. This rebound increase in blood pressure during withdrawal explains the finding that bingeing is significantly more likely to result in a stroke.[23]

It also appears that moderate drinking may not be a risk factor, for blood pressure at least. A review in *The Lancet* in

2017 concluded that if you drink three or fewer units (24g of alcohol) per day, you'll see no significant effect on blood pressure from drinking less.[24] However, anyone who drinks above that amount will see benefits if they cut down. And the more you drink, the greater the benefits of cutting down.

BRAIN

This may shock you, but drinking causes brain damage. In fact, the leading preventable causes of dementia are head injury and the damage alcohol does to the brain.

It's thought at least one in five cases of dementia is probably due to alcohol. And so a lot of people will be affected: as you probably know, dementia is one area of medicine where there are a growing number suffering. In fact, the number of people living with it is expected to triple by 2050. And it's been suggested that women may experience brain damage at lower levels of alcohol intake than men.

In a 30-year study – the Whitehall II Imaging Study[25] – which involved over 500 people having their brains scanned and their drinking histories taken, there were increased odds of hippocampal atrophy in those drinking up to 14 units weekly, and faster cognitive decline in those drinking up to 7 units weekly. The conclusion? Alcohol consumption, even at moderate levels, is associated with adverse brain outcomes.

As always, there is contradictory evidence, from a cohort study that looked at the drinking habits of 9,000 mainly middle-aged civil servants, and followed them for 23 years. It showed that low levels of alcohol consumption – that is between 1 and 14 units a week – reduced the risk of dementia. In fact, it appears that being teetotal may raise your risk of dementia – though drinking more than 14 units does too. However, the authors do say that you shouldn't start drinking if you don't already, due to all the other health harms.[26]

What is absolutely certain is that heavy drinking affects the brain. A couple of years ago, there was a man in our local neurology outpatient clinic being assessed for dementia because he was forgetful, agitated and confused. He had been tried on various, pretty expensive, medications for dementia but nothing seemed to work, and they were struggling to know what to do with him. Then, one of my trainee psychiatrists took over his case. He realised nobody had ever taken a drinking history. It turned out this man was drinking huge amounts every night, so he was in alcohol withdrawal every day. When we helped him stop drinking, his symptoms got significantly better.

There are several reasons for the impact of alcohol on the brain. There's the direct neurotoxic effect of the alcohol itself. There are indirect effects, such as the increased risk of stroke, diabetes and high blood pressure that comes from drinking. All of these raise your risk of dementia. Finally, you're more likely to have an accident after drinking – and head injury is a huge risk factor for dementia.

There's a famous condition that revolutionised our understanding of what happens to the brain in drinkers: Korsakoff's syndrome. The main symptom of this is losing the ability to lay down new memories. I've had patients come into my clinic who I've spoken to at length, and then the next time they come in, they have no recollection of ever seeing me before – even though they can recall distant memories, from their childhood or where they went to sea or fought in one of the world wars. People with this syndrome can often be misdiagnosed as play-acting. It's usually found in very heavy drinkers, and it's caused by a combination of excess alcohol and vitamin B1 (thiamine) deficiency damaging a key part of the brain that's vital for laying down new memories.

There is also an intermediate stage on the way to this, called Wernicke's encephalopathy. This is an acute inflammation of the brain characterised by a state of complete confusion – delirium, double vision, being very unsteady when standing. At this stage, if you treat the patient with thiamine (vitamin B1), you can prevent them ending up with brain damage.

That's why, for almost a century, some doctors have been saying that all alcoholic drinks should have added vitamin B1. I'd support this. Although people with acutely inflamed brains are relatively rare, it's likely there are lot of people walking around with milder forms. That is why when anyone seeks help with their drinking, we give them high doses of vitamin B1 – up to 1g per day.

WHAT IS THE FRENCH PARADOX?

The term the French Paradox was coined in 1980. It described the fact that French people had lower levels of heart disease than other nations while consuming a high level of saturated fat – think Camembert and cassoulet – and as much alcohol. (NB: the French didn't have a lower level of liver damage or cirrhosis).

The theory was, the missing link was red wine. In other words, that consuming this allowed the French to eat cheese and meat while keeping their arteries clear. The idea was that red wine was cardio-protective, but that spirits and beer were not. One chemical that occurs naturally in wine, resveratrol, came under particular scrutiny. But its early promise of being a magic cardio-protective bullet didn't come to fruition.

There have been big claims made about what resveratrol can do, including preventing both cardiovascular disease and cancer. That's why you'll see it on sale in health food stores. But most of the claims are based on *in vitro* and animal studies. And the amount you'd need to take to see any of these benefits, for the most part, is far larger than the amount you'd get in a glass of red wine.[27]

The real reason for the French Paradox is likely not wine, but the mainly Mediterranean diet the French were consuming, which was rich in olive oil and contained more fish and chicken than red meat as well as lots of fruits and vegetables, wholegrains and legumes. And this is now proven to be good for heart health. There are other factors that weren't taken into consideration besides wine: more sun and so more vitamin D, as well as the more relaxed way that southern Europeans tend to eat. There's also the fact that the French, along with other southern Europeans drink differently from northern Europeans. They're less likely to binge, more likely to drink consistently (NB: this kind of drinking with no days off may be worse for the liver).

Fig 2. The more you drink, the more diseases you get: cause specific relative risk by alcohol consumption

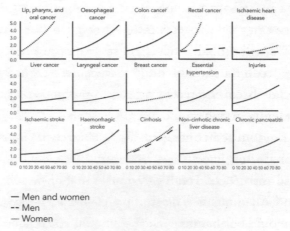

— Men and women
-- Men
···· Women

Source: https://www.bmj.com/content/325/7357/191/related

This figure shows the impact of increasing levels of drinking on the risk of a range of different medical disorders on both men and women.

Evidence does suggest that drinking small amounts of red wine might possibly be a little cardio-protective. If you look at the figure you will see that in the top right-hand box, there is a slight downward trend for ischaemic heart disease in men. You'll also see it is the only downward trend out of any of the conditions. But the optimal amount of alcohol to consume per week to get this protective effect is around one unit/day, equivalent to just 75ml of 13.5% ABV wine (less than half a medium 175ml glass in a pub). Drink any more, and the effects become negative. When consumed in that volume (or lack of it) you're really treating red wine more like having a shot of medicine, rather than a recreational drink.

No one has done an experiment that has proved definitively

that drinking alcohol – red wine or any other form – is protective for cardiovascular health. All the claims are by association, not causation, and apply to the specific population studied. So we can't say for certain that any given individual will benefit. We will probably never get definitive proof – it's very expensive to do the kind of study that could establish this.

There has been one recent attempt to set up such a study. The Moderate Alcohol and Cardiovascular Health Trial, a six-year randomised controlled trial of nearly 8,000 over-50s in 16 global sites, had its extensive funding raised by the US National Institutes of Health's National Institute on Alcohol Abuse and Alcoholism. The trial proposed that one group of subjects would be abstinent, the other would have one drink a day. This would finally answer the question: what effect does a low level of alcohol have on cardiovascular health – the rate of heart attacks, strokes and death?

The problem was, the cost – over $100 million. The drinks industry agreed to provide funding, which was in theory ringfenced from the actual study itself. But then all hell broke loose. There were those who objected because the study wouldn't include analysis of some possible negative effects, including breast cancer. But the main objection was that funding by the alcohol industry would skew the findings towards alcohol having a positive effect. Some believed drinks companies had had too much contact with the scientists involved. And there were those who thought that the researchers shouldn't take money from companies that made it by harming people. In the end, after an investigation by the US senators who regulate NIH funding, the study was dropped.[28]

DIABETES

Like cardiovascular health, there has been some evidence to suggest that drinking lightly or moderately may protect against type 2 diabetes. But also in a similar way, there is debate: a large-scale review from 2015 suggested that any beneficial effects may have been overstated and that if they do exist, they may affect only women and non-Asian populations.[29] While a 2017 review concluded that light to moderate drinking does reduce the risk of diabetes.[30]

What doctors do agree on is that heavy drinking increases the risk of type 2 diabetes.

Alcohol inflames the pancreas, the main function of which is to secrete digestive enzymes into the small intestine. These are surprisingly powerful, and what seems to happen is that once the pancreas is damaged by alcohol, it starts to digest itself, which is very painful.

This effect is dose-related, and usually happens after 10 to 20 years of drinking. Women seem to be more vulnerable than men. We know that alcohol is the most common reason why people get pancreatitis. And once the pancreas is damaged, this can lead to insulin-dependent diabetes.

A second factor that increases the likelihood of diabetes is the calorific content of alcohol. Drinking a lot increases your total calorific load, and being overweight can push you in the direction of diabetes.

If you already have diabetes – either type 1 or 2 – alcohol can affect you in other ways. Alcohol disrupts your sugar balance and metabolism, leaving you more vulnerable to both hypoglycaemia and hyperglycaemia (low and high blood sugar levels). And as alcohol affects your brain function, you might not notice what's going on or be less likely to maintain your proper insulin regime, or take steps to sort out your blood sugar.

Another big concern is that a hypo, a state that occurs when blood sugar levels drop below those needed to keep your brain working properly, can look like being drunk, so you may not get the right treatment. The thing to do here is to wear a warning bracelet, so people know you have diabetes.

THE GUT

One very common – though not life-threatening – side effect of drinking is gastric reflux. This is when the sphincter between your oesophagus and stomach opens when you lie down at night, which isn't supposed to happen. The result is that stomach acid comes up into your oesophagus and burns it, which can be painful. It's especially common if you are partial to a pint, due to the large fluid intake. But it's also due to the fact that alcohol delays gastric emptying and can stimulate gastric acid secretion. Drinking neat spirits will also burn and inflame the oesophagus.

The stomach itself seems to be relatively resistant to alcohol, presumably due to the coating that stops it being destroyed by stomach acid. There is an exception: a stomach ulcer. These are caused by bacteria called helicobacter, and alcohol makes it easier for the bacteria to bed in. Ulcers are not generally life-threatening, unless they perforate, but they can turn malignant and so lead to stomach cancer.

Alcohol is also known to impair the absorption of nutrients in the small intestine. This is especially true of vitamin B1, the malabsorption of which leads to brain damage (as discussed above). It may also affect uptake or absorption of folic acid, thiamine, B12, zinc and selenium.

Some alcoholics, those who can no longer look after themselves or who buy alcohol instead of food, or who live on the street, are vitamin-deficient in any case, due to their very poor diet. I have in the past described their diet as being just chips, which may sound flippant but it can often be true. If a patient these days presents with scurvy, which is caused by vitamin C deficiency, they are usually an alcoholic.

Drinking makes the protective membrane of the gut wall more permeable too – a condition called leaky gut. Leaky gut also allows toxic substances produced by the gut bacteria into the blood and this can lead to brain toxicity. Normally the liver mops these up but in liver failure it can't do this adequately and a state called hepatic encephalopathy ensues. One of my patients in the UK in the late 1970s was a woman dying of end stage cirrhosis - it wasn't due to alcohol but from hepatitis B. Over the years it had destroyed her liver, as alcohol can too.

She'd become confused and delirious and the reason was that the toxic substances from her gut bacteria weren't being cleared by the liver so they got into her brain and poisoned it.

We showed this was the case when we treated her with antibiotics. They kill all the bacteria in the gut, all the microbiome, so the toxins disappear. For a few days, she woke up, was alert and conscious. But then her gut recolonised itself and she went back into a coma and died.

Right now, the microbiome is a hot area of research. The suggestion is that your unique inner world of bugs is not only affected by what you drink (and eat), but is a key factor in whether you get ill and with which condition. Drinking alcohol (and/or having a bad diet) encourages the microbes in the gut to produce toxic substances that promote the inflammatory processes and gut permeability that allow diseases, including liver disease and cancer, to happen. This means the gut may also turn out to be where we can look to treat or prevent these conditions.[31]

3

AAA: ALCOHOL, ACCIDENTS AND AGGRESSION

IN THE UK, A&E departments on weekend nights are testimony to the damaging effects of alcohol. The medical staff dread the night shifts, especially today when it seems to me to be more acceptable to abuse healthcare staff than it was when I was a junior doctor. In my memory, most of my drunk patients were good-natured, even apologetic. Now they seem to be much more frequently aggressive and violent. The trade union Unison estimates that assaults on hospital staff rose 9.7 per cent between 2015/6 and 2016/7, going up to just over 200 a day.[1]

On top of chaos in the emergency room, according to an NHS estimate in 2017/18, there were 338,000 thousand hospital admissions in the UK where the main reason for hospital treatment was attributable to alcohol. Nearly a quarter

(23 per cent) were for unintentional injuries, or in other words, accidents.

ACCIDENTS DO HAPPEN

And they happen more frequently when you're drunk. Have you ever woken up from a big night out with a bruise and you don't know how you got it? Aka an unidentified drinking accident?

That's because at low doses, as well as making you clumsy, alcohol luckily has a pain-reducing effect, which seems to be mediated through turning on parts of the GABA system.

Then as you push the dose up and it affects the glutamate system, that's when the anaesthesia effect kicks in (it's also when the amnesia kicks in too, which is why you can't remember how you got that bruise/break/bump). It's a not-uncommon sight in A&E departments to see people who've broken bones and are in massive pain but are still standing. Although they may not be able to remember, or talk sense.

IN FACT, PEOPLE HAVE ACCIDENTS AND DIE

People do a lot of stupid things when they are drunk. It's a lethal combination: alcohol helps you shed your fear but also your judgement. You think: oh I can reach that – when obviously you can't. When drunk, people are also more likely to succumb to dares and challenges. I was going to write 'stupid dares' but the truth is, all dares when you are drunk are stupid.

Think of all the holidaymakers we hear about, every year, who fall and die in Benidorm and Magaluf, trying to jump into swimming pools from heights, or climb from balcony to balcony. Death and injures are so common in these Spanish resorts – especially among Brits – that this ridiculous activity has been given its own verb: 'Balconing'.

In the UK, people die from climbing buildings and falling off them. Or they die from jumping into the sea from piers and drowning – if the water is cold, your heart can stop as you hit it.

DRINK DRIVING

When I describe a car in the hands of a drunk person, I liken it to a massive weapon. Each year, around 250 people are killed in drink-drive accidents in the UK. Looking beyond the statistic, every one of these accidents is a personal tragedy for someone's family and friends. Sometimes four or five young people all die in one crash because the driver is drunk.

That's not including the many thousands of accidents where people don't die. In 2017, the latest year for which figures are available, the Department for Transport estimated that 8,600 people were injured or killed on Britain's roads in cases where the driver was over the alcohol limit. The total number of drink-drive accidents was 5,700.[2]

You won't be surprised to hear that young adults (aged 16 to 24) are over-represented in drink-drive casualties, making up a quarter of them (compared to one in five of all accidents). You may also be unsurprised that four out of five drink-drive accidents involve male drivers. Looking at the list of famous people who've been arrested for drink-driving, it is overwhelmingly male: Chris Tarrant, Wayne Rooney, Mel Gibson; and the most written about recently, of course, Ant McPartlin, half of Ant & Dec. He received a 20-month ban and was fined £86,000. At the time of the incident, his blood alcohol was more than double the legal limit. This isn't to shame him. Being dependent on alcohol is a brain disease, not a failure of will. Compounding the problem is that men tend to overestimate their capacity to drive at the best of times; and this overconfidence is made much worse when they are drunk.[3]

And let's not forget that a lot of pedestrian accidents are caused by being drunk too – because people walk in front of cars. Out in Oxford one night when I was starting my research career, I saw a man staggering down the pavement. Suddenly he walked into the road – and got hit by a car. He bounced off the car and fell over. Then he got off up and staggered away, ignoring the driver's offer to check if he was OK. He was so drunk, he didn't feel the pain.

There are three reasons drink is really bad for your driving.

1. It affects your judgement and makes you think you are a better driver than you are. Because alcohol takes away the higher-level control of your conscious brain, you lose insight into what is really going on, as well as your ability to make a rational decision.
2. Your coordination is much, much worse. Under the influence, some people drive faster and some slower. The latter can happen when drivers know they are drunk and try to overcompensate. If you have ever followed someone home

after pub closing time and they've been driving at 20mph, you can probably assume they have had too much to drink.

Your reaction times are also slower. So if something happens – a cat runs out onto the road, or another drunk person – you're not as quick at responding.

3. You're more likely to fall asleep at the wheel. Alcohol can make you much more sleepy, especially when you're tired or you're driving at night.

DO YOU KNOW YOUR LIMITS?

In one UK survey, only 9 per cent of Brits knew they needed to be under 80mg per 100ml of blood (often shortened to 80mg per cent or 80mg%). That contrasts with most European countries, where at least half of people know the national blood alcohol threshold for driving.[4]

This is what the law says: to be legal to drive in England, Northern Ireland and Wales, your blood alcohol level must be less than 80mg%. This amount was introduced in 1967 with the Road Safety Act, which is when it became an offence to drive, attempt to drive or be in charge of a motor vehicle on a road or other public place with a BAC (blood alcohol concentration) that exceeds this. It hasn't changed in 50 years, leaving the UK (excluding Scotland) with one of the highest drink-driving limits in the world; in most of Europe the level is set at 50mg% and in a number of countries, for example Norway and Sweden, at 20mg%, effectively zero alcohol intake.

In 1981, the introduction of an accurate breathalyser meant police could test drivers by the side of the road. The maximum breath alcohol concentration is 35 micrograms of alcohol in 100ml of breath (this is the equivalent of 80mg% in your blood).

If you've ever been breathalysed, you'll know what the form is: the police give you a preliminary breath test at the side of the road. If you are over the limit, you will be taken to a police station for an evidential breath, blood or urine test. If you are convicted of drink-driving, you will almost certainly be banned from driving and be given an unlimited fine. In extreme cases you may even be sent to prison.

HOW MANY DRINKS CAN I HAVE AND DRIVE?

Do you know what the 80mg% limit means for you? According to the government, it's impossible to say exactly how many drinks will take you to the limit, as it's different for each person. It will depend on your weight, age, sex and metabolism, as well as what you were drinking, and what you have eaten.

The problem is, people tend to assume they know how much takes them over the limit. Some – very sensibly – decide not to drink at all when they drive.

The rule of thumb is that an average-sized man can drink up to four units of alcohol, and a woman two (units are explained in Chapter Nine), before they exceed the UK legal limit. However, who is average? Unless you have actually been breath-alysed, it's extremely hard to know how fast your body elimi-nates alcohol – and you'll never know exactly how fast you're processing it on any one particular day.

I suspect most people over-guesstimate how much they can drink. You could work out how much you can drink according to the fact that you process one unit of alcohol an hour; but this, I think, makes for confusion. Not only do you first have to count the units in your drink but then you also have to count your drinks from the time you started, and then count the time after you stop drinking. The possible margin of error is wide, not least because alcohol affects your ability to pay attention to these details. Along the same lines, people who estimate their blood alcohol as lower than it is tend to drive more riskily.[5]

There are various unit calculators online, where you input your sex, weight and age and so on, but they all state very clearly that they aren't accurate, and that they should only be used the morning after.

I was told by one man that he always drank three pints and drove because he had, and I quote, a 'fast metabolism'. And by a woman that two glasses of wine is fine if it's with dinner. But plugging this into a blood alcohol calculator, for example, if she drinks two 175ml glasses of 13 per cent wine, it will take six hours to leave her system completely.[6]

Fig 3. The effect of drinking on risk of a road traffic accident

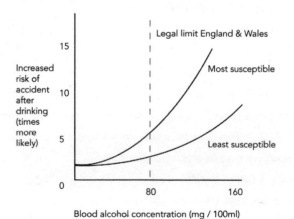

Source: www.ncbi.nlm.nih.gov/pubmed/18955613

The point is that the 80mg% is a legal threshold – it's not a safe level. The figure shows that the risks of an accident start to rise almost from when you start drinking, that any level of alcohol can impair driving to some extent. At the 80mg% limit, on average your likelihood of having a road traffic accident is at least three times that if you had stayed sober. For those most susceptible, for example young people, the risk may be up to five times greater.

And having alcohol in your blood has an even greater impact on whether you die as a result of a road crash.[7]

'How many drinks can I have and drive?' is not the question you should be asking yourself. Because it's not about staying within the legal limit, it's about staying safe.

I'd advise you, if you are driving, not to drink at all. This table, which shows how blood alcohol concentration affects the likelihood of you having an accident, might convince you:

BLOOD LEVEL/RISK OF ACCIDENT
These figures are for young people under 19 – for older people the risk rates do come down quite significantly. But 50mg% is significantly lower-risk than the current 80mg% for all age ranges. Also note the extremely steep rise in risk once blood levels get over 100mg% – this applies to people of all ages.
20mg% was twice your risk
50mg% was 3 times the risk
80mg% was 8 times the risk
100mg% was more than 12 times the risk[8]

What we are not told – when we carefully measure out our drink or drinks that we think take us up to the drink-driving limit – is that driving with a blood alcohol concentration of 80mg% is risky.

Even in 1967 – more than 50 years ago – when the level was first set, experts knew that people driving at this level were under 'significant impairment'. The figure 80mg% was only chosen as an 'acceptable compromise' at the time, when it was felt the public was against any limit at all, despite the high number of road traffic accidents at the time.

The current limit gives the impression that it is safe to drink a glass or two, but that is not the case. If being 'significantly impaired' is not a reason not to touch a drop before you drive, I don't know what is.

The *Report of the Review of Drink and Drug Driving Law* by Sir Peter North was commissioned by the UK government and published in 2010. It is a comprehensive review of drink driving and it states that if you are driving and your blood alcohol concentration is between 50mg and 80mg, you are six times more likely to die in a collision than if it was zero. And you are 2–2.5 times more likely to be involved in an accident than drivers who've drunk no alcohol.[9]

The most important message is, being under the threshold doesn't mean you're fine to drive. The difference between 79mg% and 81mg% is not that one is safe and the other dangerous, but that one is legal and the other is not.

New neuroscience research from Sussex University backs this up. It shows that even one pint of beer can compromise driving safety. Even when your behaviour and coordination are not affected, after just two units your feeling of being in control of your actions is exaggerated. And the resulting overconfidence in your driving ability could be dangerous.[10]

In many ways, a zero limit is the most rational approach (although I suspect the legislation would never get through); at zero, there is no temptation to guess or play the system.

SHOULD THE DRINK-DRIVE LIMIT GO DOWN?

In 2014, in response to the fact that Scotland had so many alcohol-related road accidents, the Scottish Parliament lowered the drink-drive limit from 80mg% to 50mg% for blood (which is 35mcg down to 22mcg per 100ml of breath).

This puts Scotland in line with almost every other country in Europe, except Malta. Germany, Italy, Austria, Croatia, Belgium, Greece, Ireland, Portugal, Spain and France are all 50mg. Estonia, Poland, Sweden and Norway are all 20mg% – which in practical terms means you cannot drink at all. The tiny amount takes account of the fact that some foods contain small amounts of alcohol, and that the body and the gut bacteria produce alcohol from food. The Czech Republic, Hungary, Romania and Slovakia have a limit of zero[11] (even though in reality, as explained above, it can never actually be zero.)

I'm all in favour of taking the limit down from 80mg% to 50mg% for the rest of the UK too. The evidence of the impact of this change in Scotland hasn't yet come through. But according to a recent report, other countries that have reduced their drink-driving limit by the same amount have seen significant falls in road accidents and fatalities.[12]

It's estimated that in the UK, the change would save between 40 and 300 lives every year, as well as hundreds of serious injuries.

Reducing the limit to 50mg% would reduce the accident risk by half, even though it wouldn't halve the amount people can drink. At 50mg% the risk of an accident for the average person is only slightly increased from having nothing to drink. Proportionally, we'd gain a lot more in terms of reducing risk than we'd lose in terms of drinking.

Again, though, for young drivers even this lower amount isn't safe. Some countries insist that learners and those who have just passed their driving tests avoid alcohol completely; they risk losing their licence if they test positive for any alcohol.

There has been some preliminary data from Scotland that's not in line with what has happened when other European countries lowered the legal limit. A study by the University of Glasgow, published in *The Lancet* in 2018, compared the rate of accidents in England and Wales with those in Scotland, and there was no reduction, according to Jim Lewsey, Professor of Medical Statistics at Glasgow University's Institute of Health and Wellbeing. He said the reduction in the drink-driving limit 'simply did not have the intended effect of reducing RTAs'. He also said this went against most other international evidence – and that the most likely explanation is that the changes weren't backed up by enforcement by police.[13]

I'd add that at the moment, in order to breathalyse you, the police need to suspect that you are drunk. But in other countries the police can do random testing, which probably acts as a more powerful deterrent to drinking.

I am also in favour of in-car breathalysers, otherwise known as alcohol ignition interlock devices. These are fitted on the cars of convicted drink-drivers in some other countries; they don't allow the car to start at all if the driver is over the limit. There is a plan to introduce them in all new cars sold in Europe by 2025. Eventually this approach might significantly reduce drink-driving accidents.

Another vital way to reduce road accidents from alcohol is to educate people that taking other sedative drugs with alcohol adds to the risks, as shown by the large Europe-wide study, Driving Under The Influence of Drugs, Alcohol and Medicines in Europe (DRUID).[14] This collected blood from thousands of people who had been involved in traffic accidents and measured their levels of a range of common prescription and recreational drugs. One of its conclusions was that combined consumption is 'most dangerous'.

In the USA, the limit is 80mg% unless you are under 21, when it is less than 20mg% – i.e. effectively zero. The limit for commercial drivers is lower because they drive more and the harms they would cause from having an accident are so much greater. Some states in the US still have functional testing, similar to how we tested drivers in the UK pre-breathalyser. This involves making the suspect prove that they are not intoxicated by having them perform a number of psychomotor coordination and memory tasks. These include toe-to-heel walking in a straight line, following a moving finger and recalling spoken sentences, all of which abilities are impaired by alcohol. If you fail the test, you can still be prosecuted even if you are under the breathalyser limit.

You could argue functional testing is fairer as it shows if you are capable of driving – or not. Although even if you are physically coordinated, your risk-taking faculties might be seriously impaired. The other issue with functional testing is that it doesn't pick up whether you might fall asleep at the wheel.

7 REASONS NOT TO DRINK AND DRIVE

1 It's much harder to concentrate and focus on one task.

2 You need to see – and drinking can give you blurred vision, bad night vision and/or even double vision.

3 Messages about what you're seeing take longer to reach your brain.

4 You are more confused about what you see, so it takes your brain longer to reach a decision to stop or turn to avoid a crash.

5 Instructions from your brain take longer to reach your body and you find coordination harder.

6 You lose the ability to judge distances.

7 You are overconfident and also more likely to take risks.

8 You may feel more sleepy and be more likely to fall asleep.

THE DEATH TRAP?

Any amount of alcohol in your system increases your risk of having an accident. And if you do have an accident, the toxicity of alcohol makes it harder for doctors to treat you. It becomes even more dangerous if you have taken any kind of sedative drug such as Valium, as the effect of the two substances add together.

My local hospital – St Mary's in Paddington – has long campaigned to reduce the harms of alcohol by early intervention, based on their findings that over 15 per cent of patients admitted to the resuscitation room of their A&E department had a positive blood alcohol concentration over 10mg%. If you look at the longest column in the figure below, what it shows is that about one third of people admitted to the resuscitation room and scoring positive for alcohol had blood alcohol levels of 240mg% (three times the drink-drive limit) or more.[15]

Fig 4. Blood alcohol levels in A&E patients

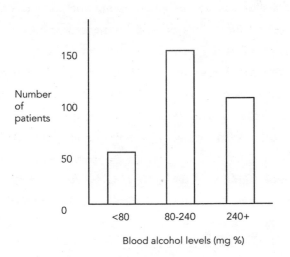

Source: www.ncbi.nlm.nih.gov/pubmed/20497803

DRIVING THE MORNING AFTER

Did you know you can still be drunk after you've been to bed and woken up the next day? In my experience, very often people don't think about the fact that alcohol takes a while to process. I'd imagine it was a shock for Sky Sports presenter Kirsty Gallacher when she was pulled over at 11 a.m. one day in 2017 and failed a breathalyser test. She'd reportedly been out until 3 a.m. and, eight hours later, her breath alcohol level was still around three times the drink-drive limit.

You don't have to have drunk an enormous amount to still be illegal the next day. The calculator at www.morning-after. org.uk shows you how long you should wait for all the alcohol to leave your system.

If you drink one 250ml glass of 13 per cent wine, for example, you won't be allowed to drive (legally) for four and a half hours.

And if you drink three 175ml glasses of 13 per cent wine, you'll have to wait eight hours to be under the limit.

That said, even if you are legal to drive the next day, you may not be safe. A review by the University of Bath was very clear on this: numerous studies show that being hungover affects key skills you need for driving, including coordination, memory and attention.[16]

WOULD A HIGHER LEGAL DRINKING AGE SAVE LIVES?

Probably the most remarkable public health intervention in terms of reducing deaths on the road happened in the USA in the 1980s. It was the increase in the drinking age, from 18, 19 or 20 – it varied by state – up to the age of 21. The US Department of Transportation forced each state to do this by threatening that if they didn't comply, they would lose 10 per cent of the federal highway building budget (a lot of money).

It is controversial, but it's been claimed that this law – the National Minimum Drinking Age Act – may have saved hundreds of thousands of lives. Young people's driving is much more impaired by a given level of alcohol, and in the vast majority of states this new law has massively reduced drink-driving deaths on the roads among under-21s.[17]

Despite this remarkable data from the USA I'm not sure I think raising the minimum age of drinking is a good idea. If someone can vote, if they can join the army, fight and die, get married, then they should be allowed to buy alcohol.

ALCOHOL AND VIOLENCE

Fig 5. Alcohol and violence

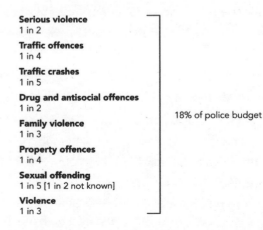

Serious violence
1 in 2

Traffic offences
1 in 4

Traffic crashes
1 in 5

Drug and antisocial offences
1 in 2

Family violence
1 in 3

Property offences
1 in 4

Sexual offending
1 in 5 [1 in 2 not known]

Violence
1 in 3

18% of police budget

Source: New Zealand Police 2010; www.alcohol.org.nz/research-resources/nz-statis-tics/alcohol-and-crime. Accessed October 2014

Alcohol makes people violent in the same way that petrol makes a fire worse. Children fight because they don't have the words to sort things out. You'd imagine adults would have grown out of that. But human beings – unlike most animals – don't need a reason to attack our own species. At least not when alcohol is part of the picture.

AUTHOR ALEKSANDR SOLZHENITSYN DESCRIBES IN *THE GULAG ARCHIPELAGO* how, during the Second World War, the Russian army would ply disposable troops with drink before an advance (these were the men whose guns were made of wood). Once drunk, they'd be told to walk towards the German lines. The idea was that the Germans would use up their machine gun bullets on them and, once that had happened, the Russian troops with real guns would be sent in to kill. The men who were ammo fodder probably wouldn't have been able to do this if they hadn't been drunk.[18]

Much further back in history, the Vikings got high on a combination of alcohol in the form of mead and magic mushrooms before their berserker raids. This was supposed to give them magical powers. It would definitely have made them seem immune to pain.[19] One theory says that the binge-drinking culture found in modern-day northern Europe reflects the spread of Viking genes throughout this region.[20]

Back in the present day, a WHO report states that in 2016, violence where alcohol is involved caused 90,000 deaths globally a year.[21] A recent UK report said there were close to 500,000 incidents of violent crime in England and Wales in 2016/7 where the victims thought the offender was under the influence of alcohol; this is around 40 per cent of all violent crime.[22]

Those numbers include sexual violence and studies have shown that alcohol is, in fact, the most common date rape drug. Despite headlines about GHB and women having their drinks spiked, a study of cases from a London toxicology lab showed that alcohol – often in high amounts – was involved in nearly half of cases (46 per cent), drugs in only a third. In only 2 per cent of cases was a sedative drug detected.[23] This is not to put any blame on the women involved: of course these crimes should not happen.

What is true is that alcohol makes you – whether you're a man or a woman – more vulnerable to becoming a victim. You're not as aware of the threats as you would be when sober and are less able to defend yourself if you are attacked.

How people transform when drunk – from mild-mannered Jekyll to violent Hyde – can be shocking. Some people who seem perfectly sane, normal and loving when sober, commit the most violent and terrible acts when drunk. You may remember the newspaper reports from 2012 when Andrew Hall, a Stoke City youth team footballer, then aged 18, stabbed his 15-year-old girlfriend Megan-Leigh Peat over sixty times, which his legal defence described as a 'catastrophic loss of control'.[24]

There are a few possible explanations as to why people change in this way. People who've had a violent upbringing tend to perpetuate that in later life. At the level of psychoanalysis you could say that the person is unhappy with him or herself and that their violence is simply them projecting that unhappiness onto others.

When people black out, they can do things – some extremely violent – that they later profoundly regret. Some people can't even remember what they did.

There was an interesting but horrific case I was involved in at the Old Bailey. A man stabbed his flatmate to death, with a screwdriver in the eyes, 50 to 80 times. He pleaded not guilty to murder. His defence was that he did it during a blackout. His blood alcohol level was extremely high; high enough, in my opinion, to make him black out. But the court decided it was murder. Being in a blackout is not a defence because the person has made the choice to drink.

One way we can think about alcohol-induced violence is to divide it into two types: Defensive and Aggressive.

1) DEFENSIVE This is violence as an unintentional reflex. It's when someone bumps into you (assuming you're the one who gets violent), spills your drink or 'looks' at you in a certain way. It is groups of men, smartly dressed for a day's racing in suit and tie at Ascot or Goodwood, erupting into a free-for-all fist fight.[25] And it's one reason sporting venues now use plastic glasses. (As an aside, a champagne bottle is one of the most dangerous weapons of all, as when it hits someone's skull it won't break).

AAA: ALCOHOL, ACCIDENTS AND AGGRESSION

One recent study from Bristol University showed that drunk people misinterpret other people's gazes. Instead of being able to read an expression of, for example, disgust, they see it as anger. This misinterpretation effect was strongest with men rating male faces.[26]

When people feel they're being threatened or having their space invaded, they then overreact to the perceived threat. This happens because alcohol has provoked a failure of the higher control centres of the brain; and when you can't control yourself, you overreact. Being drunk releases a primitive, defensive aggression that you'd normally have intellectual control over. Your frontal cortex is no longer working optimally, so you are unable to put in place your usual rational approach to what you now see as provoking behaviour.

2) AGGRESSIVE The classic example of this is a football fan – or a group of them – planning to fight fans of the opposing team after the match. This kind of aggression is intentional, done with self-control, and is about letting loose. It's done because the person wants to fight. Drinkers use alcohol for Dutch courage and drive, as well as for calories and energy.

Alcohol works for this because it deadens both fear and pain through its effects on the brain's glutamate and GABA systems, as well as endorphins. Plus the dopamine surge increases energy and drive.

ALCOHOL AND DOMESTIC VIOLENCE

It's an incredibly difficult situation to be in a relationship where one partner – most often a man – becomes violent after drinking.

One study looked at domestic abuse statistics in Lancashire during three World Cup tournaments when England were playing. The number of incidents rose by 38 per cent when England lost compared to when they weren't playing. It went up by 26 per cent if they won or drew.[27] It's not unlikely alcohol was involved in most of these cases.

Estimates of how many people who commit domestic abuse who were drunk at the time range from a quarter to three-quarters. But as studies have shown, the more severe the violence – including rape – the more likely it is that alcohol is involved.[28]

If someone is already violent, they tend to get more violent when they are drunk. If your partner has previously been violent when drunk, it shows they have the capacity and it's likely to happen again if they drink. And that is despite how sorry or regretful they are afterwards.

This may sound like old-fashioned advice, but if you are in a relationship like this, try not to have an argument when the violent person is drunk. Don't take on a drunk. They behave in ways they can't control.

Should you leave your partner? Of course you should never stay in a dangerous or violent situation if you can possibly help it. However, I have met men who have managed to stop drinking and remain abstinent, for example by going to AA, and have not been violent again.

Whether your partner is abusive or not, if you have concerns about their drinking talk about it with them when they are sober.

And if your partner is abusive, please get help.

For advice, contact the charities listed in the Resources section at the back of the book (see page 265).

WHAT ALCOHOL DOES TO YOUR MENTAL HEALTH

THERE'S NO DENYING that alcohol makes you feel good – in the short term. Part of the reason for this is that it can numb out negative emotions while you are drunk. But it also destabilises your brain chemistry.

As you withdraw from the effects of alcohol on your brain, you're likely to get psychological symptoms such as anxiety, sadness, lack of sleep and a short attention span. That's especially true after a night where you've been impulsive or behaved with a lack of judgement, aka drinker's regret. In fact, anxiety the day after drinking is so common, it even has its own nickname – hangxiety or, in Australia, boozanoia. You might get a 4 a.m. wake-up and/or an accompanying panic attack. Repeated withdrawals produce effects on the brain that, long term, add up and change your moods and thoughts, raising your risk of both depression and anxiety.

The relationship between alcohol and mental health is a complex one. Patients often present with both alcohol problems and mental health problems, so it's bit of a chicken and egg situation: which comes first? Are you using alcohol as a form of self-medication? Or has alcohol changed your brain and caused or worsened a mental health condition? There's another conundrum: as a doctor, which one should you treat first?

Probably the most common mental health issue linked with drinking is anxiety, followed by depression. But people also drink to push down other difficult feelings, such as the traumatic flashbacks of PTSD or, often, during a manic phase of bipolar disorder. We look at these conditions in more detail below.

WHAT TREATMENT LOOKS LIKE NOW

It used to be that the treatment of alcoholism and addiction came under the umbrella of mental health services. But then in 2010 the Cameron government said it didn't believe that addiction was an illness, so addiction care was moved from the NHS into social services.

The net result was that the treatment budget was halved, so services had to be cut back severely and many senior psychiatrists with years of experience in addiction were made redundant.

Now we have a real shortage of treatment places and train-ing facilities for healthcare providers in addiction. The most disadvantaged group of patients are those with complex mixtures of addiction and mental health issues, who are now largely looked after – if you can call it that – in hospital A&E departments or in prisons. The recent massive rise in deaths from opioids is likely due to this change in policy and result-ing reduction in access to treatment. I am firmly of the belief that alcoholism is a disorder of the brain, like other mental health issues, and its treatment should be embedded in psychiatry.

The problem now is that, when people present with both alcohol and mental health issues, they get pushed between mental health services and addiction services, with experts in each saying they should get the other problem treated first. In my clinical practice, I saw this passing of the buck all the time.

Alcoholics Anonymous (AA) (more of which later) is an excellent organisation but it shouldn't be the only help avail-able. The reality is, people do best with expert medical and psychological support: medication, therapy and so on. This is especially true when there are mental health conditions in the mix too.

The evidence as to how alcohol or mental health conditions affect each other isn't clear in the literature. One interesting study from Norway looked at the association between anxi-ety, depression and the amount people drank. This was a large-scale population study of nearly 40,000 people.[1] This study found that, when compared with moderate drinkers, non-drinkers had a higher risk of anxiety and of depression. It doesn't prove, however, that non-drinkers have more mental health problems; rather, it shows an association. It may be that some people with mental health conditions don't drink as a result. This study also found that heavy drinkers were more likely to have both anxiety and depression.

Conversely, a study from Hong Kong showed that, especially for moderate female drinkers, giving up alcohol will increase your mental well-being.[2] Those who quit alcohol had, after four years of abstinence, levels that approached that of lifetime teetotallers.

These are the mental health conditions that most often occur along with alcoholism.

ANXIETY

It's not surprising social anxiety is common: in evolutionary terms, it is useful. Its biological function is to protect you from strangers who might harm you. Although it's unlikely to feel particularly useful when you're standing outside a party because you can't bear to go in. Most people have some degree of social anxiety, and social anxiety disorder is simply an extreme form of this.

It was while I was working as a consultant at my specialist anxiety clinic in Bristol in the 1990s, that I realised how intertwined anxiety disorders and alcohol are.

I was called to do a home visit for a patient who was agoraphobic and so too anxious to come into clinic. He was in his 40s but he looked 80. He had peripheral neuropathy – no feeling below his knees – due to his drinking.

We started to chat about his problems. He told me that in order to go outside the house, even just to cut the grass, he needed to drink two cans of Special Brew.

I asked him, have you told anyone about your alcoholism? 'Yes,' he said, 'I told my doctor. And my doctor said I needed to go to AA. So I went to AA, but I needed four cans of Special Brew to go into the meeting. And the truth is, the people there are not like me, because they're not anxious like me.'

He was right: you won't be able to access a group therapy programme like AA if you can't speak to the other people there.

Looking at the results of one major US study of the co-occurrence of social anxiety disorder and alcoholism, we can estimate that up to a quarter of young alcoholics have social anxiety disorder.[3] In male alcoholics it's the most common condition diagnosed other than alcoholism. And the Anxiety and Depression Association of America has reported that 20 per cent of people with social anxiety also have alcoholism.[4] It's a very well established diagnosis for men, but there isn't so much comparable data for women.

It's also a perfect example of a mental health vicious circle. You start as someone who's very anxious about a social situation. So you drink to get over it, self-medicate. Each time you do this, your tolerance builds. At some point, one day as you go into withdrawal, you begin to feel even more anxious because the alcohol has changed your brain chemistry.

If we scan the brains of people affected in this way, we can see changes to the amygdala, the fear and stress centre. These adaptive changes are epigenetic – that is, they are changes at the level of gene expression; fear- and stress-promoting genes are being over-expressed.

A similar process can happen with other kinds of anxiety-related phobias, for example agoraphobia. The end result?

You are addicted to alcohol and are therefore unable to function without it.

There is another patient I remember vividly, when I was a trainee psychiatrist in Oxford in the 1980s. He was in his late twenties, and presented to our anxiety clinic with severe collapsing panic attacks. Each time he had one, he'd think he was having a heart attack and would be rushed to hospital.

His story was that from the age of 17, he'd been going to the pub. He felt anxious about talking to anyone, but the thought of talking to girls was the worst. So he drank to deal with his shyness. And that was how he socialised, every night. And his brain got tolerant to this regular alcohol intake. Then one night, after ten years of doing this, just as he was going into the pub he had his first massive panic attack.

What had happened was, over the course of the day he'd gone into alcohol withdrawal. The brain chemicals involved in this made him feel anxious. At the same time, in anticipation of the alcohol he was about to order, his brain was turning off his GABA system, which increased his anxiety levels even more, a sort of neurotransmitter double whammy. So his anxiety levels rose and before he could douse his anxiety with alcohol it reached a crescendo and he panicked.

The challenge for every doctor when faced with a case of anxiety and alcohol dependence is: do you treat the drinking or the anxiety? The alcohol specialist will say, you've got to stop drinking. But the patient can't stop drinking as he's too anxious. The anxiety specialist will say, I can't treat you as you are drinking. So he doesn't get treated. Both therapists blame the alcohol.

My reasoning is this: if we'd stopped him drinking, he would have got more anxious and wouldn't have been able to socialise. What we did – and what we do with people like him now – is treat the anxiety with either a talking therapy (such as CBT) or an SSRI medication. Eventually, this enabled him to cut down his drinking to normal levels.

So how can you tell if your drinking is due to social anxiety? One clue is if you drink before you leave the house to reduce your anxiety about going out.

Another is if you're always first to the bar. A friend who had had alcohol problems, when he was finally sorted and sober, said to me: 'You never noticed that I was always in the bar before everyone else arrived. I did that because I knew I had to have two drinks before the rest of you came, or I would be too anxious.'

It's worth saying that there are some people with anxiety who cannot bear to drink. Those people – who often suffer from panic attacks – are using all their cognitive capacity to suppress their anxious reactions. They feel that if they drink, they won't be able to do this and will become more anxious.

DEPRESSION

You may have heard that alcohol is a depressant. But do you know what that means? It means that regular, chronic alcohol use affects the serotonin system, disrupting the brain in the direction of having low mood.

Just as with anxiety, a doctor's dilemma when depression and drinking coexist is knowing which is the primary problem.

This might seem like a bit of a pointless question because we know drinking drives down mood and low mood can cause more drinking. But it is clinically important: one of the major predictors for having a poor response to being treated for drinking problems is pre-existing low mood. Another issue is that, if you drink while taking antidepressants, they won't work as well.

Alcohol is very often one of the substances that depressed people use to self-medicate. I have often said to patients: if you get depressed after drinking, be very careful. If you feel high when drinking and then low when hungover, be careful. These are both risk factors for becoming alcoholic, as they can be a driver to getting you into a cycle of drinking. If this sounds like you, I recommend being aware of this risk, and using the strategies in Chapter Nine to avoid becoming dependent. If the mood problems continue in the absence of drinking, you may need treatment, perhaps a mood stabiliser or antidepressant. See your doctor and discuss it with them.

PTSD

People suffering from PTSD often turn to drink to suppress memories and flashbacks and in order to get to sleep. Just like with anxiety disorders, self-medicating for PTSD with alcohol is a sticking plaster that can lead to alcoholism.[5]

Despite the high profile of the condition, there is still very little research into the most effective means of preventing or treating PTSD. The usual approach of debriefing has been shown to be without effect and there isn't yet a drug with enough evidence. This may be because most drug treatments are initiated well after the trauma, which may be too late to prevent the laying down of immutable brain traces for the memories, behaviours and effect that trauma causes and which develop into PTSD. Which leads me to an interesting paradox. Some studies have found that if you are drunk when you are traumatised, you're less likely to get PTSD in the first place. That's because although you may (or may not) be able to remember the facts of what happened, you are too drunk to lay down emotional memories.[6]

SUICIDE

People who have committed suicide often have high blood alcohol levels. And close to half of all people who commit suicide are intoxicated. This could be due to needing 'Dutch courage' to go through with it, or it could be that they attempt suicide because they are disinhibited by alcohol.

Also, we know that when people are drunk, they are more likely to attempt suicide and also to use more effective methods. According to WHO's *Global Status Report on Alcohol and Health 2018,* you have a seven-times-increased risk for a suicide attempt soon after drinking alcohol, and this increases to 37 times after heavy use of alcohol.[7]

In addition, being an alcoholic means you're more likely to try to commit suicide. One study found that the most dangerous time of all is at the end of a binge or the start of the withdrawal period.[8]

WHAT ALCOHOL DOES TO YOUR MENTAL HEALTH

Alcohol intoxication is also associated with attempted suicide – now called deliberate self-harm (DSH) and discussed below – again through a process of alcohol-induced brain disinhibition and despair. Preventing suicide and self-harm is very difficult, but for people with underlying depression antidepressant medications can be very effective and for those where the primary driver is alcohol, stopping drinking through medication or psychological treatments will be helpful.

OTHER MENTAL HEALTH CONDITIONS AND ALCOHOL
ADHD/IMPULSIVITY

People who are very impulsive – they may have been diagnosed with ADHD or an impulse control disorder – may be more vulnerable to abusing alcohol. For example, studies suggest that the likelihood of becoming alcoholic for an adult diagnosed with ADHD is five to ten times higher than in the general population.[9]

ADHD is thought to occur because the top-down control of behaviour from the brain's frontal cortex isn't sufficient to control the bottom-up drives and impulses. This failure of control may be because the dopamine and noradrenaline systems in the frontal cortex haven't developed adequately. That's why the most effective treatments for ADHD are medicines that enhance dopamine function e.g. amphetamines or Ritalin (methylphenidate) or those that increase noradrenaline activity e.g. Strattera (atomoxetine). The other main treatment is cognitive behavioural therapy (CBT). This helps the person strengthen their frontal control regions and also learn ways to limit their urges.

OCD

People diagnosed with OCD – repeated checking in the face of knowledge – tend to be less likely to use drugs and/or alcohol. But some do use alcohol, in order to decrease the anxiety that exists within OCD. And if they do use it, they are more likely to become addicted.[10] That's because the rituals of OCD become attached to the rituals of drinking. In fact the relationship is so strong that there are addiction experts who believe that repeated alcohol use is for many people – even those who haven't been diagnosed with OCD – more a form of compulsion than one of craving, and a scale to measure this has been developed – the Obsessive Compulsive Drinking Scale.[11]

As with the other psychiatric disorders that occur with alcoholism, treatment requires addressing both problems. For OCD, antidepressant medicines such as SSRIs (selective serotonin reuptake inhibitors) are the first-line treatment, along with CBT. It is possible that, as with the anxiety disorders, treating the OCD can help reduce the dependence on alcohol.

BULIMIA

Bulimia is an eating disorder that involves binge-eating followed by fasting, making yourself vomit or purging. There's an interesting theory, which started in the 1980s, that bulimia is the female equivalent of alcohol dependency, that they are two sides of the same coin.[12] And it's true that a lot of people who are bulimic abuse both alcohol and drugs. In one study, half of the subjects with bulimia were also alcohol-dependent.[13,14]

SELF-HARM

Drinking increases the risk of self-harm – injuring yourself. A 2005 review of 10,000 cases of self-harm showed that women tended to drink excessively particularly around the time they self-harmed (also see the section on suicide, page 100).[15]

BIPOLAR DISORDER

People often present with both bipolar disorder – having manic and depressive episodes – and alcohol abuse. Some people drink more during the high of manic episodes, some self-medicate for their feelings of anxiety. Studies suggest it may be that one triggers the other.[16]

PARANOIA AND PSYCHOSIS

It's not uncommon for very heavy drinkers to end up developing paranoia or a psychosis-like state that can make them very suspicious of other people. This is down to changes in the brain, caused by heavy drinking over time, that are similar to those seen in people with schizophrenia and which involve the dopamine system. Although rare in the general population, this phenomenon is well known to psychiatric services as it can present as very extreme and often is mistaken for paranoid schizophrenia. Both illnesses respond to treatment with antipsychotic drugs but, of course, to keep alcohol psychosis at bay it's critical to cut down on drinking too.

If any of the above symptoms sound familiar, don't ignore them. Can you talk about them with your partner or a friend? Before going to see your GP, keep a diary for a month, including your mood or eating behaviours or obsessive checking – whatever your symptoms are – along with your drinking record. That will maximise the value of your consultation with the doctor and help them work out the best way they can help you. If you have suicidal thoughts, seek help immediately.

5

HORMONES AND FERTILITY

WE TEND TO use the word hormones as shorthand for things like ovulation, pregnancy and periods but, as you probably know, they're much more complex than that. The endocrine system is a whole network of brain and nerve cells and secretory glands producing chemical messengers that interact with each other and affect all of the body and the brain. The effects happen well below our consciousness but we all experience them when we are stressed, angry, in love, hungry, tired and so on.

Hormones set the background level of most functions in the body. They are the most important homeostatic regulators, keeping our body running smoothly hour to hour and day to day. So if they go wrong, it can lead to real health problems.

For instance, as well as reproductive hormones, there are thyroid hormones, growth hormones and stress hormones. Each of these hormones can either directly change how our body functions or they can turn on other hormones – or both. All of these hormones don't just come out of the brain and the glands where they're produced, they feed back into them too, switching themselves and each other on and off. It's a system that's always trying to get itself back into balance; when you add in alcohol, it can quickly get out of whack.

Take what happens when you go into withdrawal from alcohol after a heavy night out. The main system where hormones are produced and interact is called the hypothalamic-pituitary-adrenal (HPA) axis. These three glands produce hormones that control the stress response. This happens when your brain or your body gets stressed – which includes when you are coming down from alcohol, aka withdrawal.

There are two parts to the stress response.

1) There's a stress pathway direct from the brain to the adrenal glands, which sit on top of the kidneys. The adrenal glands release the stress hormones adrenaline and noradrenaline. And these are the hormones that make your heart pump faster, make you feel alert and hyperactive and wake you up at 4 a.m., sweating and nervous.

2) The brain turns on the hypothalamus, which releases a hormone called CRF (corticotrophin-releasing hormone). This travels in the blood down a special set of veins to the pituitary gland. This is the master gland of the endocrine system and controls most hormones. It sits just under your

brain, in the middle of the head at the back of your eyes. When the CRF hits the pituitary gland, it stimulates the release of another hormone called ACTH (adreno-cortico-trophic-hormone, in case you're interested). ACTH travels through the bloodstream to the adrenals and prompts them to release the stress hormone, cortisol.

The noradrenaline/adrenaline and cortisol systems work together to provide a coordinated response to stress. The first part of the stress response happens immediately; then, over the next minutes and hours, cortisol rises. If you drink every day, and thus turn on that system repeatedly every morning as you go into mini withdrawal, you're pushing yourself towards high blood pressure. You always have to have some cortisol in your blood to keep you alive but if it gets too high, tissue damage occurs. Too much cortisol is like taking steroids: it alters your ability to metabolise sugar, so it also pushes you to diabetes. And, over time, it can also drive changes in the brain that can lead to and perpetuate depression.

WHAT ALCOHOL DOES TO SEX HORMONES

Prolactin is the hormone that promotes milk production. And it's been shown that levels are higher in heavy drinkers of both sexes.

In women, prolactin also has a role in regulating ovulation. Even a single dose of alcohol can increase prolactin levels, although it is dose-dependent. One study of women showed that a single episode of drinking moderate to high amounts of alcohol – around five units – made prolactin rise and oxytocin (the bonding hormone) go down.[1] The effect is thought to be transitory, but if you drink every day, levels will stay high. And one theory says that high levels of prolactin may be behind the increased risk of breast cancer that comes from drinking.

Men need low levels of prolactin for sperm production. But if levels rise, you can get breast growth and testosterone levels go down. This is one reason men who drink heavily end up being feminised. Alcohol also lowers testosterone via other mechanisms; it increases levels of an enzyme that breaks it down and turns it into oestrogen. This can happen even before alcohol starts to cause serious liver problems. And acetaldehyde (produced in the liver when alcohol is broken down) impairs the production of an enzyme that's key for testosterone synthesis.

Because waste hormones are broken down in the liver, once liver disease begins a man will not be able to metabolise oestrogens as well as before. So he will become even more feminised, possibly grow breasts and lose his facial hair. The feminisation is also partly due to a drop in IGF1 – insulin derived growth factor – which leads to less muscle, less testosterone and less sperm. Male alcoholics also lose muscle mass. Besides not eating properly, one of the reasons is that they don't manufacture enough HGH (human growth hormone).

DOES DRINKING REDUCE FERTILITY?

The short answer is, yes. Alcohol causes the production of destructive elements called free radicals, which can damage proteins and DNA. You may have heard of free radicals already, in terms of sunlight causing free radical production in the skin. Although it's normal for your body to produce free radicals, the addition of alcohol means too many are produced and in the wrong place, so they then harm the body. It's thought many of the toxic effects of alcohol in the liver, but also in the ovaries and testes, are down to free radicals. This process is called oxidative stress because the body's proteins – including DNA – are oxidised (damaged).[2]

Experts agree heavy drinking does decrease male fertility. There are obvious ways: it lowers libido and increases the likelihood of impotence too. Then there is its effect on sperm. The jury is out on whether a (very) moderate amount of alcohol decreases or increases sperm quality and volume.[3] One study showed a higher sperm count for moderate drinkers (in this case around three to five pints of beer a week) than those who drank less.[4]

But another study showed that drinking any more than five units a week (2.5 pints of beer) affected sperm quality and quantity. And the more men drank, the bigger the effect.[5,6]

There's a similarly mixed picture for female fertility – although the general message is the same: that alcohol probably does damage fertility. One study showed fewer than 14 units a week doesn't seem to have a measurable effect on female fertility.[7]

However, a review of studies of nearly 100,000 women showed that drinking at any level reduced fertility – 13 per cent (for any drinking), 11 per cent (for light drinking: around 1.5 units or a very small glass of wine a day), and 23 per cent (for moderate-to-heavy drinking: more than a small glass of wine a day).[8]

In another study, 50 per cent of social drinkers (people who drink up to four drinks a day) and 60 per cent of heavy drinkers (four to eight drinks a day) had disruption in their menstrual cycle and altered sex hormones. Half of the social drinkers who had more than three drinks a day had anovulatory cycles – that is, they didn't ovulate.[9]

CAN I DRINK WHILE I'M PREGNANT?

British women drink more during pregnancy than their counterparts in ten other European countries, according to a 2017 study. In the study 28.5 per cent of women in the UK drank during pregnancy, followed by 26.5 per cent in Russia. In Norway, it's just 4.1 per cent who drink during pregnancy. One reason for the high levels in the UK is that, in the past, guidelines have varied and so the message of whether it's safe to drink has been unclear.[10]

The guidelines are now clear. NHS guidance states: 'Experts are still unsure exactly how much – if any – alcohol is completely safe for you to have while you're pregnant, so the safest approach is not to drink at all while you're expecting.'[11]

This is because evidence suggests alcohol increases the risk of stillbirth, premature labour and foetal growth restriction. Babies of drinking mothers also run the risk of withdrawal straight after they are born – they can even experience the DTs (delirium tremens) if they are not diagnosed and treated.[12]

Drinking also increases the risk of foetal alcohol spectrum disorders (FASD), a range of mental and physical problems caused by exposure to alcohol during pregnancy. As a poison, alcohol crosses the placenta and goes into the developing foetus, potentially causing a huge range of lifelong issues.[13]

The most easily diagnosed type of FASD is called foetal alcohol syndrome, in which the baby has specific and identifiable facial features: a smooth philtrum, thin upper lip, small eyes, low-set ears, small head size. But these children, it's thought, only account for around one in ten children who have FASD. The obvious physical changes are due to the mother drinking during weeks six to nine, when the facial features are forming. But it is the other aspects of FASD – caused by drinking at different times during the pregnancy – that are possibly the biggest burden. These include disturbed and hyperactive behaviour, low intelligence, developmental delays, impaired memory, problems with hearing and vision, growth impairments and problems with the heart, kidneys and bones.[14]

There's some evidence that bingeing during pregnancy is worse than a more spread-out pattern of drinking. No level of consumption is safe, but the risk seems to go up particularly at around 28g – around three to four units per week.[15,16]

The global prevalence of FASD is estimated at around 8 in 1,000 in the general population but it can be very much higher in certain high-drinking populations. So it's not a small issue: there are many more affected kids than those with, for example, Down's syndrome, which is 1 per 1000.[17] The country with the highest rate of FASD is South Africa, where more than one in ten children have it. One reason might be that in some rural areas in South Africa, the vineries pay workers partly in alcohol. This practice was made illegal in 1961 but has been reported to exist still.[18,19,20]

One big issue for parents is that the brain damage part of FASD will happen during the first trimester. That's why the recommendations are clear not to drink if you are planning to be pregnant. If you discover you're pregnant and you've been drinking, stop. And try not to worry. The reality is, for the vast majority of women, the risk to the child is going to be very low. It's estimated that 1 in every 13 pregnant women who drinks while pregnant will have a child with FASD. And the risk is strongly related to level of consumption.[21]

Alcohol isn't all bad news for pregnancy. It was one of the first drugs used to delay premature labour, though the evidence that it helps isn't that good and it's since been replaced by other medications.[22]

HOW DOES ALCOHOL AFFECT THE MENOPAUSE?

It's not clear whether drinking has an effect on the timing of the menopause: some studies seem to show it makes it earlier. Others seem to suggest low to moderate drinking might be associated with a later menopause.[23]

What has been shown is that alcohol can make the menopausal symptoms of sweating and hot flushes worse, because of its effect on the HPA axis and the resulting increase in noradrenaline.[24]

If you have any of the depression or anxiety symptoms that can go along with menopause, you might find drinking makes them worse, too. Menopause makes some women more vulnerable to these because the changes in hormone levels affect neurotransmitters such as serotonin and noradrenaline. These same hormone changes after the menopause may alter alcohol sensitivity, so you may find the same amount of alcohol gets you more drunk. Also, both men and women get more sensitive to alcohol as they get older due to a reduction in body fluid which results in a higher blood alcohol concentration.

Another worry that women often have at this time is the risk of breast cancer. Your risk goes up with age and, as you already know, alcohol further raises your risk as does taking HRT. In fact, there is likely a cumulative effect on breast cancer risk if you both drink alcohol and HRT. This means the risk goes up more than if you added the two risks together.[25]

WHY DO I GET MORE DRUNK SOME TIMES OF THE MONTH THAN OTHERS?

Surges and dips in progesterone and oestrogen – which happen naturally over the course of your menstrual cycle – change the sensitivity of your brain to alcohol. Progesterone, just like alcohol, works on the GABA system in the brain. Add alcohol to a progesterone surge, and you'll likely feel more drunk.

During pregnancy, it is the super-high levels of progesterone needed to keep the pregnancy that account for memory problems, aka baby brain. High progesterone levels, in effect, sedate you. You could say you are pre-drunk.

Most of the breakdown of progesterone happens in the liver, although it also takes place in the brain, skin and other areas. Progesterone can be broken down via two pathways. The pathway will depend on the balance of some specific enzymes in your body:

1. In some women, progesterone breaks down to a sedating, anticonvulsive chemical called allopregnanolone. What is really fascinating is that this might help explain post-natal depression. During pregnancy your brain becomes

dependent on the high levels of progesterone and adapts to them in a way similar to how your brain becomes tolerant to alcohol. Then, after birth when progesterone levels drop, you go into withdrawal. Now a new drug that's a synthetic form of allopregnanolone – brexanolone – has just been licensed to treat post-natal depression. It is given intravenously for several days to restore the post-natal deficit.

2. The other breakdown route progesterone can take is to pregnenolone sulphate. The result of this is exactly the opposite to that of allopregnanolone. It decreases the effects of GABA so causes anxiety and seizures. It feels like alcohol withdrawal. This explains why, as progesterone fluctuates during your cycle, and depending on what else is happening in your body, you may become either more anxious or relaxed.

Your current oestrogen levels also affect how alcohol affects you. When oestrogen is high, alcohol gets you higher, due to its effect on dopamine receptors. This increased euphoria and pleasure can put you at higher risk of addiction.[26]

It's likely that the hormonal fluctuation of the perimenopause years are the reason why women find their relationship with alcohol changes, whether it's that they can no longer tolerate it or they want to drink more.

CAN I DRINK AND BREASTFEED?

You may have heard the old wives' tale that stout is good for breastfeeding. Although drinking does increase prolactin, beer's effect on milk production is more likely to be due to the barley extract it contains – which explains why non-alcoholic beer has been shown to work just as well.

And if you were wondering how much alcohol gets into breast milk, a study has shown that it is 5–6 per cent of the dose the mother receives, a very low amount.[27]

That's why NHS advice is that if you are breastfeeding, the limit is the same as if you aren't – 14 units a week. If you do drink 14 units in a week, spread them out over a few days. And don't feed for two to three hours after alcohol.

GUT HORMONES AND ALCOHOL DRINKING

The stomach, which doctors used to think of as just a bag that held strong acid and enzymes to digest food, is now known to produce hormones. One of these is called ghrelin (a strange name, but it's based on its role as a *growth hormone-releasing peptide*). Its function is to make you hungry. But it can also make you thirsty; it's been shown to induce craving for alcohol as well as food.[28]

Ten years ago, an influential paper showed that this system determined alcohol liking in mice; when injected into mice, it increased their alcohol intake. And when its effects were blocked, the effects of alcohol in the mice were also blocked.[29]

Since then there have been several studies in humans showing similar effects.[30]

What we've seen so far implies that stopping ghrelin production or blocking its receptor actions would reduce drinking. My research team and I have been studying this in our Medical Research Council-funded study, GHADD, run by Dr Tony Goldstone, using an anti-ghrelin hormone called desacyl-ghrelin. To check this out we are brain-imaging abstinent alcoholics while exposing them to drinking cues. Normally these cues activate brain regions involved in craving alcohol, but he's shown that when desacyl-ghrelin is given, this activation reduces, which we hope may mean it will reduce drinking. Unfortunately desacyl-ghrelin needs to be injected, so it can't easily be used as a treatment. Research is under way, however, to find orally active versions. Once these are available, we plan to test them.

6

HOW ALCOHOL AFFECTS YOUR QUALITY OF LIFE

YEARS BEFORE ANY health issues show up, alcohol is already affecting every organ and system of your body, including your brain. And there are more subtle changes happening too. It's affecting your health, yes, but also your wellness: your performance at work, your skin, your fitness, your sex life.

In this chapter, I hope to answer the questions: is how much I drink adding to my life, or taking away from it? What's the level of drinking that will make me feel good but allow me to function well too?

ALCOHOL AND SLEEP

After a night out, do you tend to suddenly wake up in the early hours, feeling alert and awake? What you are experiencing is a mini withdrawal from alcohol. This hyperactive, fight or flight state is prompted by a surge of the neurotransmitter noradrenaline. You may also be sweating, and you may find it hard to get back to sleep.

What can you do to stop this? You'll need to work out, through trial and error, the limit that affects your sleep. Research suggests you'd have to drink three units or more to disrupt your sleep significantly. But firstly, three units is only a pint and a half of 5 per cent beer, or a glass and a half of wine. And secondly, your limit will be personal to you.

If you keep drinking enough to go into withdrawal every morning, the stress response will accumulate. Then, as well as the spike in noradrenaline, your cortisol levels will go up. This alters your ability to metabolise sugar and can push you towards diabetes or a rise in blood pressure. Cortisol may also drive changes in the brain that underlie depression.

Some people get into a habit of drinking to fall asleep. And it works: alcohol puts you to sleep faster and into a deeper sleep more quickly too. It wasn't uncommon, when I was first working on a hospital ward in the 1970s, for doctors to prescribe a glass of sweet sherry to elderly patients at bedtime.

The problem is, in the second part of the night your sleep becomes more disturbed. You may not get as much REM (Rapid Eye Movement) sleep, which is when you dream, and which helps the brain consolidate memories.

If you are using alcohol to sleep, be careful: you're increasing your risk of dependence. The best way to improve your sleep is to improve your sleep hygiene. For example, exercise during the day, in the morning and ideally outside to get sunlight, both of which help reset body sleep–wake rhythms.

Wind down for bed without any screens, and sleep in a cool, dark room. You could also try cognitive behavioural therapy; there are some good introductions to it online, such as the Sleepio app.[1] This is free from GPs in some parts of the country; but even if you have to pay, the cost will probably be offset by your savings on alcohol.

If neither of these strategies work, do see your GP. A short course of sleeping aid might help you reset your sleep rhythm. If the insomnia persists, ask your GP for a referral to a sleep expert.

Do you snore? If so, you've probably been told that you're much noisier after a drink. Snoring is a sign of the beginning of an obstruction in the air going in and out of your lungs. Drinking makes it worse because alcohol relaxes your muscles, including those of your pharynx and palate, the soft tissues at the back of your mouth and in your throat.

Snoring can develop into sleep apnoea: this is where your airways collapse so you can't get any air into or out of your lungs. It deprives your brain of oxygen, a condition called hypoxia, just as if you had a plastic bag over your head.

A hypoxic episode can last 20 or 30 seconds until your brain gets so stressed that it fires off a massive fight or flight response. You then take an abrupt breath and your airways open for a few minutes, until you go back into a deeper level of sleep. Usually the stress response doesn't wake up the person having it, but the snoring and spluttering that accompany it can be very irritating – and possibly worrying – to any bed partner, who may feel that the apnoeic person is never going to start breathing again. In my sleep clinic I would come across non-drinking partners staying awake for hours 'just in case' they had to revive their partner.

For many people, this pattern of hypoxia followed by a mini stress response repeats itself, sometimes hundreds of times a night. Over days and weeks each episode adds up, which can lead to daytime sleepiness but also problems such as hypertension and obesity.

Treatment involves keeping the airways open with a Continuous Positive Airway Pressure machine (CPAP). The downside of these machines is their noise; they tend to keep the person's poor partner awake and may be even worse than the snoring.

Both snoring and sleep apnoea are more likely if you're overweight. There's a vicious circle I used to see quite often in my sleep clinic: fat men who snore and drink and keep getting fatter. Poor sleep increases ghrelin, the hormone from the stomach that increases your appetite for food and for alcohol too. It also reduces the level of the satiety hormone leptin. (Spikes in ghrelin may be partly responsible for the fact that you feel so hungry the day after drinking, too.)[2]

A woman in her mid-twenties once came in to the sleep clinic with terrible alcohol dependency. She was drinking half a bottle of vodka every night and suffering withdrawal in the daytime. I asked her why she only drank at night and it turned out she was so anxious about going to sleep, she couldn't go to bed unless she was drunk.

The reason for her anxiety was because she suffered from sleep paralysis. This is when your brain wakes up, but your body doesn't, so you can't move and you feel as if you can't breathe. She told me it felt like someone sitting on her chest. Another patient described trying to get out of this state of paralysis as being as exhausting as 'climbing up the curtains using only my arms'. You can see why, with that kind of unpleasantness in store every night, someone might turn to alcohol to help them get to sleep.

The woman told me that she'd wake up two or three times a night, feeling as if she was dying. The irony was, although the alcohol got her to sleep, the withdrawal was making her episodes of sleep paralysis more frequent. We treated her by helping her to stop drinking, along with a short course of a benzodiazepine sleeping medicine while the SSRI medicine we also prescribed got to work. SSRIs are usually described as antidepressants but they are very good for anxiety disorders too and I have found them useful in a range of different anxiety-related sleep disorders including sleep paralysis and night terrors.

Night terrors are different from nightmares: terrors come during the very deepest stages of sleep, unlike nightmares, which come during dreaming sleep. People will often scream and shout, might even get out of bed to fight off whatever they feel is attacking or threatening them. They appear terrified but they are deeply asleep. Terrors are relatively common in children, who usually grow out of them, but adults get them too.

Alcohol makes night terrors more likely, because it pushes you into deep sleep. Night terrors are due to some parts of the brain being awake from fear, while the consciousness areas are still asleep. And alcohol helps keep them unconscious.

I once treated a 21-year-old student who came to my sleep clinic after a holiday in Europe with a university friend and his parents. When the holiday gîte was full of guests, the two students were asked to spend a night in a tent outside, in the woods. After an evening socialising over food and wine, my patient went to bed. In the middle of the night, during a night terror, he thought he was being attacked by a bear. He felt its paws on his throat so, to defend himself, he bit it. But what had really happened was, he'd been screaming, his friend had tried to wake him by putting his hands on his shoulder – and he'd almost bitten off his friend's thumb!

Like most patients with night terrors, his were worse when sleeping in new places. We often had university students coming to our clinic in their first few weeks at college when they were away from home for the first time, and drinking more too.

But the strangest sleep problem linked to alcohol I've come across was two sisters who had Sleeping Beauty syndrome, aka Kleine–Levin Syndrome (KLS). This is a rare neurological disorder where people – usually teenagers or young adults – fall asleep for three or four days at a time. We don't know what causes KLS. We think it comes from some form of inflammation – perhaps a virus – damaging the wakefulness-promoting centres of the brain. There's almost certainly a genetic component as well.

The same thing happened to each of the sisters, albeit three years apart: the very first time they drank alcohol, they fell into a deep sleep for days. Afterwards, they rarely drank again, and never to get drunk. We know from hangover research that withdrawal can lead to considerable inflammation of the brain (see Chapter Two), so perhaps this was the case for them.

ALCOHOL AND YOUR APPEARANCE

Alcohol dehydrates your whole body, but your skin is where you'll see it. In fact each unit of alcohol – that is, 8g – makes you pee out roughly 80ml extra of urine. So if you drink one 175ml glass of 13.5 per cent wine, you'll pee out nearly 200ml extra. Which is why it's a good idea to hydrate before drinking, have a glass of water with every glass of wine, and a pint before bed, too.

Dehydration not only makes your skin look less plump but it also makes dark eye bags more visible, because it dilates blood vessels (vasodilation). This is also why drinking often leads to spider veins and to rosacea, which looks like flushing and redness on the cheeks. In one study, the more women drank, the higher the risk of them developing rosacea. The worst kinds of booze for this? White wine and spirits. NB: in this study, red wine made existing rosacea worse but didn't provoke it.[3]

The body-wide inflammation provoked by alcohol will show up on your skin as puffiness. And inflammation is likely to be why some people find alcohol is a trigger for their psoriasis, too.

In terms of ageing, alcohol is bad for all the organs of the body, from the heart to the liver to the brain (see Chapter Two). So it's not surprising that, when a Danish study looked at heavy drinkers, they found that they had more outer indicators of inner ageing than people who drank less.[4] Is cutting down good for your skin? According to a survey by the University of Sussex, half of people who quit for Dry January said their skin improved.[5]

ALCOHOL AND NUTRITION

When Nigel Lawson, the ex-chancellor of the Exchequer, was asked how he managed to lose a lot of weight in a short time, his answer was that he stopped drinking. He explains why brilliantly: 'When I've had half a bottle, I forget the reason for not drinking the second half. Or a second bottle. Along with the reasons for not ordering the prawn cocktail, the roast beef, Yorkshire pudding and the apple crumble. So Chardonnay has to go, in my experience, not just for its calories (700 a bottle) but for its unintended consequences.'[6]

He's right. Alcohol takes away the higher functions of the brain and makes it more likely that you will eat uninhibitedly. (But you probably already know that.)

Let's not forget about the calories in booze, too – it is made by fermenting sugar or starch, after all. One estimate is that alcohol can make up as much as 16 per cent of adult female drinkers' total energy intake.

Alcohol calories contain minimal nutrition. I have often seen recommendations for people who want to lose weight to drink spirits with soda. But pure spirits are likely not the best regular drink in terms of health. In theory, one unit of any alcohol will have the same toxicity for the liver and other organs, but both beer and wine may possibly contain chemicals with some health benefits. You'll have heard of the claim that substances called polyphenols in red wine are good for your heart (see 'The French Paradox', page 62), as well as being protective against cancer and possibly anti-ageing too. And there's data to show that beer drinkers don't seem to die as young as spirits drinkers, possibly because of the amino acids and peptides beer contains.[7]

In terms of energy, beer is carbohydrate-rich and contains around 180 calories a pint. But studies suggest it isn't specifically responsible for the 'beer belly'.[8]

The typical hard, sticking-out domed gut is made up of visceral fat – fat between the organs – and probably some of the soft and flabby kind of fat – subcutaneous fat – too. Both are made by excess calories rather than beer itself (although, see the hormones section for the possible oestrogen-like effects of alcohol).

Drinking can also feed into a cycle of weight gain, as explained in the sleep section above. If you drink too much and don't get enough good-quality sleep, it boosts your levels of ghrelin, the hormone that increases alcohol- and food-related activity in the brain – i.e. makes you want to eat and drink more. Then you're more likely to eat junk food and, it's been shown, to drink over the guidelines. Which then affects your sleep . . . and so on.

Doctors do wonder why some people get ill from drinking, for example developing cirrhosis, and others seem to be more resilient. A review in *The Lancet* has pointed out that the health harms of alcohol fall more on drinkers of lower economic status. What this means if, you are poor and drink too much, you are more likely to get sick. And that's despite the fact that we know consumption of alcohol across the social strata in this country is pretty similar. In the analysis, the researchers show this isn't due to either smoking or obesity.[9]

Diet could be one possible contributory reason. There are obviously other problems that come with being poor: stress, bad housing, not being able to afford heating and so on. But the real downwards spiral often happens for people who drink excessively when they can't afford food. Perhaps a balanced diet provides some protection against health harms? What we have known for a long time is that giving supplements of the B vitamins thiamine (B1) and B12 can (partially) protect the brains of people who are vulnerable by reasons of poor diet. Indeed, as we saw in Chapter Two, thiamine deficiency can lead to permanent brain damage and memory loss in Korsakoff's syndrome. It's thought that this deficiency has a generally negative impact on brains that are already subject to assault by alcohol, acetaldehyde and inflammation.

If you drink regularly – and have a less than perfect diet – I'd suggest you take a multivitamin supplement that includes vitamins D and K. Or at the very least, take a B vitamin complex.

DRINK / UNITS / CALORIES

Pint of 4% beer / 2.3 units / 182 calories

Bottle (330ml) of 5% beer / 1.6 units / 142 calories

Pint of 4.5% cider / 2.6 units / 216 calories

Medium glass (175ml) of 13% wine / 2.3 units / 159 calories

Glass (125ml) of 12% champagne / 1.5 units, 89 calories

Shot of spirits (25ml) / 1 unit / 61 calories

Source: Drinkaware.co.uk

ALCOHOL AND WORK

It seems incredible now that it used to be normal – almost part of the job description – to get drunk at work, to have convivial lunches and schmooze clients. Drinking even took place in consultants' restaurants in hospitals and in senior common rooms in universities. My life as a junior hospital doctor in the 1970s was spent either on the wards or in the hospital bar. Every hospital had one consultant who would regularly be unfit for action because of being drunk. In one of my first senior house officer jobs in a top London hospital, I once found a senior professor collapsed on the floor. As I tried to get him up and into an office, the professor of medicine walked by, saw him lying there, sighed, 'Oh no – not again!' and helped me carry the incapacitated prof to his office to sleep it off.

During my first stint on a Medical Research Council grant panel in the 1990s, they served wine at lunch. Though not

everyone drank, I suspected that grant scoring was more leni-
ent in the afternoon – unfairly for the morning grants. After a
few of us pointed this out, the alcohol was discreetly removed
from the menu.

Long lunches still happen in some industries but they are
for high days and holidays, not the everyday norm. Now most
industries are dry, at least during the daytime, and in the NHS
drinking is a dismissible offence. The reason for this is obvious – alcohol impairs performance. But also, lunchtime drinking has become unfashionable, so there's social pressure
against it too.

In Japan, there's still a tradition that you seal a business
deal with alcohol, and drinking after work with your
colleagues is seen as obligatory. You must never leave the bar
before your boss, which is why workers stay overnight in tiny
hotel rooms built just for this purpose.

In China, which now consumes half of the trillion-dollar-a-
year global alcohol market, celebratory drinking at meals
with the shout of '*Kampai!*' is standard (or at least it seems to
be when I am there). I find avoiding getting drunk at every
meal the most challenging aspect of being in China, even
more so than the language barrier.

In the UK, companies now use booze as a bonding mechanism, with Friday happy hours and team drinks and, of
course, the classic Christmas party booze-up. The fallout
from those has probably led to more Dry Januaries than any
other factor.

It does seem that companies are going to have to become
more creative and less alcohol-focused with their after-work

activities since the advent of #MeToo and the fact that fewer people now drink. That includes ex-alcoholics as well as those who've chosen to go sober and people who don't drink for religious reasons, such as Muslims and Jehovah's Witnesses.

According to one US survey, one in three people (35 per cent) preferred not to drink alcohol at work events.[9] Sixteen per cent said they drank at them despite not wanting to, and 12 per cent pretended to drink in order to fit in. The biggest morning-after regrets of those who did drink? Revealing personal secrets to colleagues (14 per cent) and complaining about work-related issues to the boss or co-workers (13 per cent). Nine per cent said they embarrassed themselves by getting too drunk and eight per cent said they 'engaged in sexual activity with a co-worker.'

Which brings us to the morning after. Have you ever spent a day slumped at your desk, achieved barely anything? People reckon they are around 40 per cent less productive when hungover. New research from the Institute of Alcohol Studies (IAS) estimates that hangovers cost the British economy up to £1.4bn a year in lost productivity. And that, every day, as many as 89,000 people go to work hungover.[10]

A review of 19 studies by the University of Bath has shown that it's your sustained attention – concentration and focus – that's hardest hit by a hangover.[11] Your memory is also affected, both short- and long-term, but the problem is in making memories, rather than retrieving them.

It sounds obvious, but hangover symptoms are dose-dependent. Australian researchers – who breathalysed subjects, then tested their memories – found those whose brain function was the most impaired had had the highest blood alcohol concentrations the previous night, too.[12]

In short, if you've got a hangover, you won't be doing your best work.[13] So you might as well spend the day in bed (though it may be hard to get a doctor's sick note!).

WHAT ALCOHOL DOES TO FITNESS

You already know that moving your body is one of the best things you can do for your health: it's good for energy levels and for mental and physical health, from your brain to your heart, lungs, muscles and bones, and it protects against chronic illnesses.

What you may not know is that there's a positive relationship between alcohol and exercise: i.e. exercisers are more likely to be moderate drinkers and vice versa. What's not known is why: is it due to team sports, or a go-hard-go-home living ethos? Or maybe some people exercise because they feel guilty about drinking?

However, that doesn't mean alcohol is good for your fitness performance. Not only because of dehydration, which is more severe when you add sweating into the mix. But also because alcohol leads to fluctuations in blood sugar, a drop in muscle strength and, of course, being tired. It can also disrupt hormone levels – including the growth hormone, which you need for muscle gains.[13,14]

Despite the fact that alcohol isn't great for fitness, it's surprising how often you come across the two together. Sports clubs often use alcohol sales to underpin their finances, and a lot of people enjoy drinking after a game or match.

Drinking in football has claimed some high-profile casualties, notably Paul Gascoigne. Tony Adams is the most eloquent.[15] The ex-England captain has bared his soul about alcohol damaging all aspects of his life. He now supports other players who are having problems with addiction.

Every few weeks, a footballer seems to be in the national papers for drink-driving, the most recent at time of writing being the Chelsea and England player Danny Drinkwater, who was charged with drink-driving after a crash.[16]

When all professional sport took place on a Saturday, drinking after matches and at weekend parties could often remain hidden. But now sport is almost a 24/7 activity, it's more difficult to hide your hangover from your manager. This may be the reason why football players are turning instead to laughing gas (nitrous oxide) for a quick blast of fun.[17,18]

By inhaling a balloon of gas, they can get a high similar to that of alcohol without any risk of drink-driving or a hangover. Nitrous oxide clears from the body in a few minutes rather than in hours. And unlike other recreational drugs, such as cocaine, nitrous oxide is legal, provided a person uses their own supplies and doesn't provide it to their friends (which is illegal as it's classed as supplying drugs).

ALCOHOL AND SEX

Shakespeare had the measure of alcohol and sex. In *Macbeth*, Macduff says: 'It provokes the desire, but it takes away the performance; therefore much drink may be said to be an equivocator with lechery: it makes him, and it mars him; it sets him on, and it takes him off'.[19]

Because a lot of sex is in the mind, it might be the thought of alcohol being an aphrodisiac that is the real effect, rather than any physiological effect.[20]

In one study, 12 heterosexual men and 12 heterosexual women were interviewed about taking marijuana and/or drinking before sex. It found that although alcohol made people feel more attractive and helped them meet a partner, it was also more likely than marijuana to make people choose 'atypical' partners and to have 'post-sex regret'. Which translates as: they slept with the wrong person. Both drugs increased the risk of unsafe sex. Unsurprisingly, teenagers are more likely to have unprotected sex when drunk too.[21]

Then there is what alcohol does to sexual function. It's been shown to help women feel more comfortable about looking at sexual images – less inhibited, said the researchers – but it didn't make them feel more aroused.[22]

Women with alcohol problems are more likely to have sexual disorders, such as low libido and the inability to orgasm, although it's not clear if this effect happens with moderate drinking.[23]

As the expression 'brewer's droop' implies, alcohol and erections are opposing forces. The higher the alcohol intake, the more likely a man will have unsatisfying orgasm, low libido and/or premature ejaculation.[24]

In the five years of an Australian study that followed 810 men aged 35 to 80, 31 per cent developed some form of erectile dysfunction.[25]

As well as alcohol, other risk factors for erectile dysfunction are age, being overweight and not sleeping enough. And be aware: it's also often an early marker for cardiovascular disease. The good news is, a third of those men were able to improve their sex lives – by cutting down on alcohol, but also by stopping smoking, eating better and taking exercise.

But there are some positives to drinking, too . . .

DRINKING MAY MAKE YOU MORE CREATIVE

Hard drinking has always been associated with artists, writers and thinkers, from Jackson Pollock and Beethoven to Dorothy Parker and, further back, Socrates and the ancient Greeks. It was reported by Herodotus, the Greek historian, that the ancient Persians would only finally make a decision after the issue at hand had been discussed both sober and drunk. The idea was, alcohol liberates your thinking – but you shouldn't make a decision when you are drunk.[26]

The converse is, there is some evidence to say alcohol can help you to be creative. The psychological theory behind this is, alcohol stops you from focusing well and cuts down your working memory. And so it widens the scope of things you have in mind, and, by doing this, allows you to think out of the box.

In a study done to test this at the University of Illinois in Chicago, the participants were given vodka and cranberry juice, then asked to do a creative word game. It was the kind where you have to think of a word that links three other words (for example skate, cube and pack – the link is ice). Compared to those who did it sober, the drinkers were more accurate and faster – and they said their problem-solving felt more intuitive too.[27,28]

DRINKING AND THE ROYAL NAVY

If you can keep people drunk, you can get them to do what you want. While not exactly a positive in the sense of the rest of this section, you can see how alcohol could be a useful tool in the military. In the seventeenth century, the Royal Navy started using alcohol as a way to control their sailors. It wasn't a desirable job, so many of the men were press-ganged. There's even a story of one poor soul being kidnapped from the steps of the church on his wedding day. Rum rations were there to make the sailors' lives feel less tough as well as give them the necessary courage to scale the rigging when the ship was moving beneath them. And also, rum made them alcohol-dependent. That meant they could be controlled by the threat of having it taken away.[29]

In the early 1700s, each naval man over the age of 18 got half a pint of over 50 per cent pure rum each day. The sailors' way to prove it was of that strength was by seeing if gunpowder soaked in rum would still burn (like we might set fire to brandy or sambuca today). This is where the term 'per cent proof' came from. As a pint taken all at once made most sailors too drunk, in 1740 Admiral Edward Vernon split the ration into twice a day and decreed that it must be diluted four times with water. His nickname was 'Old Grog', which is where we get the word grog as slang for alcohol. Later on, in an attempt to prevent scurvy, lime or lemon juice was added to the ration.

Even diluted, half a pint of rum is still 24 modern units, which is undeniably a lot per day. So in 1823, the ration was cut in half, and then again in 1850. This 'tot' of rum (an eighth of a pint measure) given at noon stayed until 1970, when it was replaced by a beer ration, presumably as this was thought to be less impairing.

DRINKING AND AIM

Alcohol is a very good treatment for the shakes. That doesn't mean the shakes from withdrawal but just the normal kind of tremor we get when stressed or anxious. That's one reason darts players drink. Champion player Andy Fordham says he was never sober until he went teetotal in 2007. And he says he won the 2004 world title after most of a bottle of brandy and two dozen bottles of beer. 'For me, it helped the concentration, numbed everything, you weren't aware of what was going on behind you, you could just concentrate on what was in front of you, the board. I know it was terrible for my health, but it just seemed right.'[30]

There's also a condition where you shake, called benign essential tremor. One of the treatments is alcohol. There was an interesting court case in 2014 of an anaesthetist in France who killed a patient by intubating her oesophagus instead of her windpipe.[31,32] She had been drinking rosé wine. Part of her defence was that she had to drink in order to stop the shakes. But as she was three times over the drink-driving limit, she may have been drinking to stop the shakes of withdrawal rather than those of essential tremor. Possibly she had started out drinking to treat her benign essential tremor ... This would make her another example of alcohol initially alleviating symptoms such as anxiety or sleeplessness, for example. But then over time leading to dependence.

DRINKING AND CONFIDENCE

There's long been a theory that extroverts are more affected by alcohol than introverts. This is based on work in the 1950s by psychologist Hans Eysenck. The theory was tested in 1958, in one of the very first scientific studies that looked at the relationship between blood alcohol levels and driving performance. It found that impairments to driving start at 20mg% and got worse.[33]

In addition, the study found that extroverts who drank drove less well, but introverts were much less affected. In fact, the driving of 30 per cent of the introverts actually improved.

This feeds into what we now know about the brain. My theory is, in introverts the GABA system is set too low, so that they receive an instantaneous boost in confidence from drinking. As with social anxiety, if drinking does this to you, be careful: you are more at risk of drinking too much as the incentive to keep using it is stronger.

The confidence effect is also thought to be behind a study that showed people learning to speak Dutch became better after consuming the equivalent of a pint of 5 per cent beer; they not only spoke more fluently but their pronunciation was better too.[34]

ADDICTION: HAVE I GOT AN ALCOHOL PROBLEM?

I drank to drown my sorrows, but the dammed things learned how to swim.

Frida Kahlo

IT'S NOT YOUR fault you want to drink. Alcohol is a powerful and addictive drug that's not only legal but enmeshed in our everyday life. It's almost impossible not to come across it as we grow up. And most social occasions and celebrations involve drink.

Being both relaxing and pleasant to take, alcohol encourages habitual use. Alcoholism is common, even (in fact especially) in the medical community. There's a joke that goes: What's the definition of an alcoholic? Someone who drinks more than their doctor!

Alcohol can make you feel like the person you want to be, too. So when people say, 'Oh I adore the taste of my 1984 claret,' I find it funny. Because I know that if you drank it for a few days and it didn't get you feeling relaxed and cheerful, you wouldn't be so keen on drinking it.

In the UK, government figures from 2016 estimate that 10.8 million adults are drinking at levels that pose some risk to their health. And 1.6 million adults may have some level of alcohol dependence.

Most people drink to relax rather than to get high and so don't understand addiction. One of the reasons we as a society have to think carefully about addiction is that there's a view, popular with conservative (small c) politicians, that addiction is fun and addicts enjoy getting drunk, and that they are weak. What I've seen, for most people who are alcoholics, is that they don't want to do it; they have to do it, so much so that they can't not do it.

There are lots of factors that add up to why people become addicted, which I'll cover in this chapter, including genetics, lifestyle, life events – and the fact that it is an extremely powerful drug.

You could think about it like this: the majority of people who take heroin use it occasionally for fun. But 40 per cent of those who try it get addicted and then cannot *not* use it. For alcohol, this figure is around ten to 15 per cent.[1]

Countless addicts have told me they don't want to drink, can't stop and don't know why. I've seen people ruin their health and their life.

One of the first times the strength of addiction began to be revealed to me was when I was aged 26. I'd finished my medical training in Britain and, after travelling through India and Myanmar, I ended up in Australia doing some postgraduate training, and working every weekend as a locum for a GP. Half of my cases were alcohol-related, whether it was people diving into empty swimming pools, self-harming or having fights.

One morning, around 9 a.m., a woman called me in to see her husband. When I arrived, he was staggering around, really drunk, very confused. He kept saying, 'I can't see' and 'I don't know what's going on.' I ascertained that he'd been out drinking most of the night and had got home at 4 a.m. He'd fallen asleep on the sofa and, when he woke up, he began staggering around with blurred vision. His wife was extremely angry with him.

I thought, is he still drunk? To all appearances, he was. But when I examined him, I found significant inflammation at the back of one of his eyes. He wasn't just drunk. At some point in the night, he must have banged his head; he had a bleed on the brain that was causing pressure on his optic nerves.

At this point, in through the back door came two of his mates, with bottles of brandy in their hands, asking, 'Is Steve coming out?' His wife started shouting at them: 'No way, you're killing him.' Even with his brain injury, he wanted to go with them.

In this chapter, I'm going to explain why it's not people's fault when they become addicted to alcohol, how this happens and what you can do if this happens to you or a loved one.

WHAT IS AN ALCOHOLIC?

The Department of Health divides people and drinking into five categories. I'd say that all of category five – Dependent – are alcoholics, and at least some of category four are too.

1. Non-drinker.
2. Low-risk. Regularly consuming 14 units or less.
3. Hazardous (increasing-risk drinking). A pattern of alcohol consumption that increases someone's risk of harm. Regularly drinking more than 14 units a week but less than 35 for women or less than 50 for men. These people are probably not dependent physically, although they may be psychologically.
4. Harmful drinking (high-risk drinking). A pattern of alcohol consumption that's causing mental or physical damage. Drinking 35 units or more a week for women, 50 or more for men. A large proportion of alcohol consumption is done by people who drink very heavily – that is, over ten units a day. Not all of these people will be dependent, but many will.
5. Dependent. Typically has a strong desire to drink and difficulties controlling its use. Continues to drink despite harmful consequences.

The National Institute for Health and Care Excellence (NICE) defines harmful drinking as a pattern of alcohol consumption that causes health problems, including psychological problems such as depression, alcohol-related accidents or physical illness. It says that harmful drinkers can become alcohol-dependent, characterised by craving, tolerance, a preoccupation with alcohol and continued drinking in spite of harmful consequences.

In my definition of an alcoholic I'd include: someone who either gets into trouble when they're drunk, or who has withdrawal symptoms such as morning shakes that require alcohol to stop, or who binges when they don't want to, or who drinks secretly.

Outside the medical community, people use the term 'functioning alcoholic'. But I'd say this simply means someone whose alcohol use hasn't led to them losing their job. There's a myth that all alcoholics are lying in the gutter. But a vast number are getting by, just not functioning as well as they could. And a proportion of those people are on the slippery slope to not functioning at all.

In the past, there has been a lot of controversy over the terms alcoholic and addict. Historically, those were the words used for people addicted to alcohol. Then from around the 1960s onwards, it began to be thought that these terms were stigmatising, that they were self-defeating, that they put people into a box from which they couldn't escape.

The new terminology was 'dependence', and the condition was 'drug or alcohol dependence syndrome'. The basic idea was, if you got a physical withdrawal reaction from a substance, you were dependent. However, a lot of substances, including medicines, cause a rebound reaction when you stop them, and so technically cause dependence. Under such a broad definition, you could argue, for example, that you are dependent on antidepressants. But is that really dependence in quite the same way as alcohol?

I would argue not. I don't think this is a helpful way to describe addiction to alcohol. It doesn't take account of the psychological side, including the craving and desire.

Now we talk about both physical and psychological dependence. It's known that some substances – including alcohol – can be addictive for some people without causing physical dependence. For example, some extreme bingers will absolutely have the desire to get drunk twice a week, and won't be able to stop themselves because they are addicted to the high. But they won't go into withdrawal.

It's also known that with other drugs, such as benzodiazepines, you experience physical dependence because you go into withdrawal without them – but the difference is, you don't crave them.

My opinion now is that the idea of addiction needs to encompass both the psychological and physical. And that, when it comes to alcohol, people will have differing amounts of each of these.

So my definition of addiction is: a behavioural disorder underpinned by changes in the brain, which leads to continued use of a drug or substance in the face of problems such as withdrawal; and that your use of that substance interferes with your family and social life and causes you personal harms. And to go back to my original point, many people who are addicted to alcohol call themselves addicts and/or alcoholics. If this term is good enough for them, it's good enough for me.

HAVE I GOT AN ALCOHOL PROBLEM?

Below are the kind of questions a psychiatrist would ask you to assess if you have a problem with drinking. If you answer yes to any of them, it doesn't mean you are addicted to or dependent on alcohol but it does mean you may be moving towards that. If you think this might be you, your GP is a good person to talk to about it.

☐ Do you drink more than you intend to on a regular basis?
☐ Have people criticised you for your drinking?
☐ Has people criticising you for your drinking annoyed you?
☐ Have you got into trouble or missed work because of your drinking?
☐ Have you been ill because of your drinking?
☐ Have you ever had to drink in the morning to deal with the shakes or nausea?
☐ Do you think you're spending too much time drinking and doing less of the things you used to do?
☐ Are you spending more money on alcohol than you really should?

HOW DOES ADDICTION HAPPEN?

There are multiple stages towards addiction. And you don't need to be physically dependent – number four in this list – to be addicted.

1. People start drinking because they like the pleasure it gives them. They also like the fact it makes them feel different, gives them a buzz. A lot of people like to drink because it stops them being nervous or anxious too, a typical effect of this drug.

2. Tolerance to the effects of alcohol rises. That means your body becomes adapted to the effects, so it's prepared for it and you can drink more without being impaired – but it also means you need to drink more to get the buzz. There's no rule about how fast tolerance happens – we all vary – but it's likely to happen over a few days of heavy drinking.

3. Over time, people get more preoccupied with drinking. They might make sure more of their social life is in the pub, go out every Friday night with mates to drink, start to look forward to drinking. They seek out events where alcohol is easily available, make it a bigger part of their life.

4. As they get more tolerant to alcohol, people may start to get physically dependent (although this doesn't happen to everyone). The sign of this is that they go into withdrawal after drinking. This means they wake up early in the morning feeling shaky, anxious, nauseous, depressed. This might sound like a hangover and it's true, the symptoms are the same.

I know my tolerance develops pretty fast. When I was captain of London University's badminton team in 1974, we did a tour in Sweden. A back injury meant I couldn't compete, so I spent my time playing table tennis and drinking the vodka the Swedes had asked us to bring them from Berlin because it was so expensive in their country. One morning, after two days of drinking, I remember going to put sugar on my cereal and realising my hands were shaking. The alcohol had changed my brain; I'd developed tolerance and, as I hadn't drunk for a few hours, I was in withdrawal. So I stopped drinking for the rest of the tour. But if I had drunk that morning to stop the shakes, and then done it again the next day and the next day, I might have become dependent on alcohol.

It's when people start to drink to get rid of the symptoms of withdrawal that the downward spiral into alcohol dependence begins. If you need alcohol to keep going after you've drunk alcohol, you are physically addicted. Withdrawal is a physical phenomenon that proves you are dependent on alcohol.

Whether you get withdrawal or not, drinking can become very compulsive. In your brain, pathways of behaviour are laid down by the release of dopamine – this is true of alcohol and cocaine. The important transition is when people go from choosing to use alcohol to having a compulsion to do it.

A lot of the newer research relates to the fact that once people are addicted, breaking out of that addiction can be extremely difficult because of these changes in the brain. In my team's work at Imperial College, we have looked at brain scans of alcoholics who are abstinent and found they have altered functions in the dopamine regions of the brain. We have also shown that some other neurotransmitters, especially GABA and endorphins, are changed too.

On scans, we can see the original motivation to use a drug starts in a part of the brain that has a degree of knowledge and choice. This isn't your conscious mind; it's called the drive or motive circuit, the part that recognises alcohol's positive effects and then decides you're going to take it again.

Eventually, as the person continues to drink over weeks and months, this behaviour is downloaded into the habit part of the brain, so it becomes hard-wired. At that point, the behaviour happens subconsciously, without you making a decision to do it. Now, it's very hard for your conscious, thinking brain to get in the way.

I had one patient in the US, a man aged 27, who dried out in the clinic, then left and said he was going to stay abstinent by going to AA. A few weeks later, he was back in the clinic. I asked him, what happened? He said, 'I don't know. I found myself in a booze shop, drinking a quart of bourbon. I don't remember paying for it. I didn't want to drink it. I have no idea how I got there.'

That is an extreme example of the subconscious drive overwhelming the conscious mind. You may think it sounds unlikely, but I believe him and the other alcohol addicts who've had similar stories.

We can describe this process in terms of four brain circuits that underpin addiction:

1) The reward circuit: that is, the hit of dopamine or endorphins you get when you drink.
2) The memory circuit: how you remember that alcohol is rewarding.
3) The drive or motive circuit, the area that makes you want to eat, drink and have sex.
4) The conscious mind, the highest part of your brain. This is the part that should have top-down control. But it can be overridden: when your friends say, 'Come to the pub! It'll be fun,' your conscious mind might think, no, remember my terrible hangover last week? But your drive circuit remembers the great time you had and makes you forget your hangover and agree to go.

As you get addicted, the memory of the pleasure that's been registered by the reward circuit gets stronger. This then turns up the volume on the drive circuit, which in turn is more able to override the conscious mind.

Normally, most of us don't do things we shouldn't because our conscious mind says no. But over time, chronic drinking damages the brain so it becomes less stable, especially during withdrawal, and so leaves you with even less self-control.

TO TREAT OR PUNISH?

A hundred years ago, we used to lock people up to get them to stop drinking. The reason we don't now is because it's been shown that stopping drinking is easy; it's keeping sober that's so difficult.

There's an interesting debate currently about whether people should be forced into treatment. The argument goes: alcoholism is a mental illness, so why shouldn't we prioritise its treatment and treat people compulsorily, as we do people who have a psychotic disorder, for example schizophrenia? The justification for actively treating that group is to stop them potentially being dangerous to themselves or society. Chronic alcoholism is mostly dangerous for the alcoholic themselves – they lose 20 years off their lifespan – although it can also have a huge impact on their families and on society at large in terms of personal violence plus crime, cost to the NHS and so on.

There is one big difference: regular medication for schizo-phrenia works, but it is harder to keep people from drinking as treatments for alcoholism aren't as effective. And forcing people to be abstinent doesn't work: even in prison, for example, inmates will go to the effort of making a mush of chewed-up bread and fruit or juice, fermenting it, then drinking it.

The other reason treatment for alcoholism isn't prioritised is that society and particularly politicians prefer to view alcoholism as a failure of will rather than a failure of the brain. This kind of conflation of morality and science happens in addiction more than any other branch of medicine. It is a way of de-medicalising addiction.

The result is that politicians can justify the massive attrition in financial investment in treatment (for example, the number of consultants who specialise in this area has halved since addiction services were moved from the NHS to social care). Most treatment is now not within the NHS, although you still access it through your GP.

Another result of budget cuts is that alcohol and drug treatment resources have been pooled. This means that often everyone – regardless of age or reasons for drinking – ends up in the same treatment group. The problem with this is, there is a big cultural and socio-economic difference between drinkers and the biggest other group who use addiction services, heroin users. This presents big challenges for therapists, who ideally should individualise interventions to the patient's age, background, work opportunities, social support etc.

I would argue that although the budget for addiction services may have been reduced, it has resulted in a much bigger burden, financial and otherwise, on the police, on prisons and on the NHS, who all have to pick up the pieces.

SELF-HELP FOR OVERDRINKING

1) Be aware of your addiction and be honest with yourself about it. What most therapy aims to do is strengthen the prefrontal cortex, the part of your conscious mind that makes decisions. This is the first step.

2) Be aware of your triggers. Don't go to pubs if you don't want to drink. Remove yourself from places you associate with alcohol. Don't socialise with the people you always drink with – unless they are happy to go out to a place where you can both avoid alcohol.

3) Work out your drivers to drink. If you can identify the motives you have for drinking, you can then target your intervention and replace what you are missing or need in your life. Make a list of all the other things you can do. If you feel unable to control yourself, can you think of ways that you can rearrange your life so alcohol isn't so present? If you are stressed, it's worth trying out various ways to relax: meditation, mindfulness, self-hypnosis, relaxation and breathing techniques, yoga or just going for a walk. It doesn't matter what it is, as long as it works for you.

4) If you have low mood or anxiety, would you benefit from an antidepressant or talking therapy? Proven natural mood-lifters are: exercise, spending time outside and being in nature. If you crave excitement or a buzz, what non-drinking activities might give you that buzz (that aren't also self-destructive, such as sex with strangers or gambling)?

5) Continually remember the benefits of not drinking. This means that, when you are confronted with the desire to drink, you won't have to work to remember the drawbacks. Make a list of all the positive benefits of not overindulging, as many as you can, and keep it with you.

6) Know the times you're going to be vulnerable and most likely to relapse. Research suggests there are three of these. The first is obvious: times of stress. Typical examples of this might be divorce, a death, losing your job or work stress. If you are drinking every night to deal with the stress of work, you need to tackle the stress and/or alter the way you think about work. The second time is less obvious: times of success or celebration. People often relapse when something good happens, for example a promotion. The excitement leads to them believing they have 'cracked it' or feeling that they deserve a drink. If you think about it, success is, in effect, a positive kind of stress, because now you have to maintain this new status. The third is alterations in mood. For example, if you feel depressed or just unusually low after drinking, be very careful. Because it's easy to get into a vicious cycle of drinking to overcome depression. That goes for any kind of drinking you do to change your mood.

AM I BECOMING DEPENDENT?

There are some warning signs that you may be tipping over into dependence, meaning that the amount you are drinking has changed and is now changing your body and your brain. Treat alcohol seriously: monitor the amount you are taking, and where and when you are doing it.

The issue is that becoming dependent can lead to other problems, whether they involve relationships, work, finances, accidents, being arrested or drink-driving, to name a few. If any of the below sound like you, now is the time to take stock, to stop or cut back.

- You notice you can drink more than you used to. I'd say that's especially true if you've doubled your consumption in the past few months. Think about it like this: if you doubled your consumption of food, you'd be concerned.
- You realise you have started drinking by yourself. And especially if you are drinking alone in order to quell anxiety.
- Every night, you really look forward to drinking when you get home from work.
- You have begun to end a day's work by drinking. It's a habit rather than planned. For example, you find yourself opening a bottle or going to the pub without having thought about whether you really want a drink.
- You insist on finishing the bottle of wine. Bottles will keep, at least overnight. But you may find yourself rationalising that the wine won't be as good the next day, or it will go off. At this point, you're beginning to drink for the alcohol rather than enjoying the taste.
- You have started drinking more than your partner. For example, you might be having three-quarters of the bottle of wine, and your partner is only drinking one quarter.

Or you drink half a bottle, then open a second one just for you.

- Your partner or a friend or relative points out that you become unpleasant at some point during drinking. (NB: even if you can easily drink a bottle of wine a night without it impacting your relationships, you're still at risk of health harms and/or dependence.)
- Your partner encourages you to drink. Or you and your partner collude in drinking – i.e. you both encourage each other to open the second bottle. It's like the play *Who's Afraid Of Virginia Woolf?*, where the relationship dysfunction of the main couple is reflected in the amount they drink together. If what you do together is drink, together you'll have no resistance either. You can often see this in couples who run bars or pubs together.
- You're the one at the bar, getting in the shots. You never need shots.

I remember my shock when I first went to a Scottish pub and they were drinking pints with chasers. The more you mix your drinks, the harder it is to control the speed and amount you drink.

- You are often the first person to the bar – because you need the alcohol. It also reveals that you're the person who is prepared to make the most financial sacrifice for it.
- When the pub closes, you're the one who often says, 'Let's go to a club!' Or you tend to extend your evening of drinking by moving from place to place.
- Drinking has become your hobby or the only way you socialise. This often happens with retirees or expats.
- You have regular blackouts. A blackout means you've

drunk so much, you've switched off your conscious brain. Regular blackouts mean you've done this repeatedly and haven't learned from it.

- You regularly vomit after drinking. Vomiting is protective, as it stops you dying from alcohol poisoning. Whether you vomit relates to how fast you drink and how much alcohol is in your stomach. NB: if you vomit in your sleep, it can be lethal.

- You drink in secret. It could be taking a hip flask or half-bottles to work, maybe tucking them in a drawer. Or it could be solitary drinking at home, when other people can't see you doing it. Do you hide your bottles or empties?

- You start drinking every day at lunchtime. This has become less socially acceptable at work, but you may be in your job partly because in that company culture it's still usual or even required. It's very common to start drinking early on holiday and at weddings and celebrations. But do you carry on with this behaviour on normal days?

IS THERE SUCH THING AS AN ADDICTIVE PERSONALITY?

There is no such thing as an addictive personality. But there are individuals who are more vulnerable to becoming addicted. Your sex will make a difference, as men are more likely to be dependent on alcohol. According to the NHS, around 9 per cent of men in the UK show signs of alcohol dependence against 3 per cent of women.[2]

Whether you get addicted can depend on your brain circuitry – maybe it didn't get formed correctly, or has been damaged, either by the drugs themselves or other injuries. People who are impulsive or anxious or have another mental health condition are more likely to become addicted. Addiction also often depends on your life circumstances; as you can imagine, the most vulnerable times are those of high stress: divorce, bereavement and so on.

IS ADDICTION IN YOUR GENES?

One of the reasons I got interested in alcoholism was reading about Danish adoption studies, from the 1950s and 60s, on the children of alcoholics. It showed that these children's risk of becoming alcoholics as adults was the same whether they were brought up in a drinking family, a normal family or an abstinent family. This strongly suggests some inherited basis for alcoholism rather than it being a behaviour learned from the alcoholic father.[3]

Since then, it's been shown alcoholism does run in families. If your parent has had or has alcoholism, your own risk is three to four times higher.[4]

But you don't just inherit alcoholism; in many cases, you inherit one or more of the various different traits that make it more likely you'll be addicted, for example impulsivity or conduct disorder, as shown by a recent study on twins.[5] Some traits are mediated through a predisposition to other factors we've mentioned, for example anxiety or stress.

Others relate to the effects of alcohol on the brain. And some relate to the way your body deals with alcohol when you drink it.

One of the most important findings on vulnerability to alcoholism came from research on the sons of alcoholics by

Prof Marc Schuckit at the University of California at San Diego. It found male children of male alcoholics had significantly higher rates of alcoholism than the general population. And more detailed analysis of those boys found that they seemed to have inherited a tolerance to alcohol, possibly even from the first time they drank it.[6] This meant they were able to tolerate higher amounts than their peers. Yet despite being to some extent pre-tolerant, they were more likely to be alcoholic. It could be that they got some kind of prestige from being able to drink more, or that they just found it easy to drink more than their friends, and this made them vulnerable. Or it could be their heavier drinking produced other changes in their brain that then led to dependence and alcoholism.[7] I believe the trait being passed on, subsensitivity to alcohol, may be due to an alteration in the GABA system. The same kind of genetic factor appears to apply to women with a family history of alcoholism too, although it seems this effect is not as strong.[8]

You can also inherit a tendency to be less likely to become an alcoholic.

Probably the best known tendency for protection comes from the alcohol flushing reaction. One estimate is that 70 per cent of East Asians (Japanese, Chinese, Taiwanese and Korean ethnic groups) have this, and 5 per cent of European ethnic groups.[9]

And we know that Chinese people have lower rates of alcoholism and reduced rates of the toxic effects of alcohol, such as cirrhosis and liver cancer, too.

If you have this trait, you will have a variant of the gene aldehyde dehydrogenase (ALDH) that slows down the breakdown of the alcohol by-product acetaldehyde, which is a poison. This means that if you drink, you get high blood levels of acetaldehyde – which is what makes you flush and sweat, as well as giving you a headache, nausea and speeding up your heart rate. The alcohol treatment disulfiram (Antabuse) also blocks this enzyme, as described in the 'Treatment' section in this chapter. If you take it and drink, you get the same unpleasant effects –- the flushing reaction – which (the point is) can help you resist the temptation to have another drink.

You can get the same effects from taking certain drugs, including the antibiotic metronidazole (Flagyl), which is why it has a warning on it not to drink. And for anyone who's had a drink while taking it, I'm sure the experience has taught them one thing: to listen to their doctor!

I experienced this effect myself many years ago when, before a formal dinner for neurologists at St Mary's Hospital, I took quite a big dose of phenylbutazone (Bute) for my arthritis pain. At the time, it was quite a popular anti-rheumatic (it's still used for horses). I couldn't understand why I was so inappropriately and illogically drunk, flushed and loud, from a normal level of formal dinner drinking. My girlfriend picked me up, and said I was more outrageous than normal. I was even sick afterwards. I looked it up the next day, and discovered that phenylbutazone also blocks AIDH. It wasn't all my fault, which is probably why she agreed to forget the episode and marry me . . . Wikipedia has a good entry on the many other drugs that can block AIDH – if you're prescribed any medication, it's worth reading before you drink![10]

Each gene has two parts, or alleles. If they are both defective for AIDH, you will get the flushing reaction very badly and will almost certainly learn to avoid drink. If only one is defective, you will have just some of the enzyme and so may be able to force yourself to put up with the reaction, as it won't be so extreme. The downside is that if you do drink, you will have more toxic acetaldehyde going around your body for longer, and research shows that will put you at a higher risk for cancer, in particular oesophageal cancer.[11]

Recent advances in technology have made it possible to measure changes in all the genes in the body – this is called whole genome analysis. Analysing data from about half a million people, a team from Imperial College London found 46 new genetic markers linked to alcohol intake, which together accounted for 7 per cent of the variation in people's total drinking. The team also identified genetic pathways shared between alcohol intake and brain networks associated with psychiatric disorders such as schizophrenia.[12]

WHAT CAN LEAD TO ADDICTION

Anxiety can lead to addiction: imagine you are a young man, who gets nervous about, for example, talking to girls. When you go out, you drink because you find it reduces your anxiety. This happens because alcohol increases your levels of the neurotransmitter GABA and decreases your levels of glutamate, so you feel relaxed.

If you drink regularly, i.e. most days, then your tolerance will start to build, there will be a physical change in your brain. In response to the increase in levels of GABA, your GABA system becomes down-regulated. At the same time, your glutamate system is up-regulated. Now, you have to drink more to get the same level of anxiety reduction.

Also, you'll find that when you stop drinking, you feel even more anxious. You now have a hyper-excitable brain that needs more alcohol for you to feel normal. You reach a point where if you don't drink you are too anxious to function. You are now a dependent drinker.

Having fun can also lead to addiction; some people get a huge burst of pleasure from drinking – an endorphin rush. As your brain gets tolerant to the increased levels of GABA, it's also becoming sensitised to the increased dopamine and endorphins.

In this case, the more you drink, the bigger the buzz you get. One of the reasons people who respond like this lose control over their drinking is that as the buzz increases, they drink more. Their dopamine system gets turned on, so they continue to seek and desire alcohol and enjoy it – even as they become more tolerant to its effects. They are motivated towards drinking even if they don't want to.

Another of the reasons people turn to alcohol is because, as well as reducing physical pain, it gets rid of emotional pain – it takes away irritation, frustration and tension. In fact, it acts on the brain's emotional circuits in a similar way to opiates. Some alcoholics report that they feel numb and protected when they are drunk, that life feels like a better place. There are all kinds of reasons for this kind of pain that are beyond the scope of this book, from illness and poverty to disability and loss. It's important to remember that, for a lot of people, life is tough and the world is hard and challenging. For some, alcohol relieves this pain.

THE FAST TEST

This is one of the initial tests a GP can use to assess your drinking.

	0	1	2	3	4	Your Score
How often have you had 6 or more units if female, or 8 or more if male, on a single occasion in the last year?	Never	Less than monthly	Monthly	Weekly	Daily or almost daily	

	0	1	2	3	4	Your Score
How often during the last year have you failed to do what was normally expected from you because of your drinking?	Never	Less than monthly	Monthly	Weekly	Daily or almost daily	

	0	1	2	3	4	Your Score
How often during the last year have you been unable to remember what happened the night before because you had been drinking?	Never	Less than monthly	Monthly	Weekly	Daily or almost daily	

	0	1	2	3	4	Your Score
Has a relative or friend, doctor or other health worker been concerned about your drinking or suggested that you cut down?	No		Yes, but not in the last year		Yes, during the last year	

An overall total score of 3 or more on the first or all 4 questions is FAST positive.

Alternatively, here is an even shorter questionnaire a GP can also use:

Do you regularly have more than six drinks in one sitting? Or do you regret a drunken escapade that took place in the past year?

If the answer is yes to either, take the full FAST questionnaire.

SO YOU THINK YOU HAVE A PROBLEM . . .

You may know you have a problem with alcohol. Or you may suspect it, after reading this book. Or it may be your problem has been picked up in the GP's surgery. As the health harms of alcohol are now so well recognised, GPs are encouraged to ask questions about drinking.

The FAST questionnaire above is a selection of questions from a longer test GPs can use, called AUDIT.

If the result of your FAST/AUDIT assessment is that you

drink too much, your GP will give you advice on cutting down to sensible levels or on giving up drinking altogether. Getting your doctor's advice may be just what you need to do this; it is for many people – pioneering research by Professor Griffith Edwards, a psychiatrist and a leader of treatment research for alcohol dependence in the 1960s and 1970s, showed that firm words from a doctor can encourage people to cut down.[13] Since then there have been many studies of what has become known as a 'brief intervention' in primary care, an evidence-based, structured conversation, showing these to have real benefit. These are designed not to tell you off, but to help you see the risks of your drinking and the benefits of reducing and to motivate you to change.[14]

Your GP may also ask further questions to find out if you are dependent on alcohol.

If you try to cut down and find you can't, because of either physical or psychological symptoms, or if your GP assesses that you are dependent, the next step is to be referred to specialist alcohol services.

DETOX: WHAT HAPPENS?

When you stop drinking, you go through a detox, which simply means physical withdrawal from alcohol.

If you have tried to do this on your own but found it too hard because the cravings were too strong or you felt very ill or anxious, you will be offered medical assistance. There is a very well-established reduction protocol for detoxing, usually done at home with medication prescribed by your GP. Much less often, it is done in hospital or at a private rehab centre.

With the help of medication, you should be able to come off alcohol safely and without a great deal of distress. Medication will stop unpleasant symptoms caused by the deficiency of your GABA system, such as anxiety and visual hallucinations and, more importantly, life-threatening ones such as seizures or delirium tremens during withdrawal.

During a detox, you are transferred from alcohol to another drug, usually the benzodiazepine Valium or Librium. Then, over the course of a week, the medication is reduced slowly, allowing your brain to adapt and recover. You do lose a week and it's not exactly enjoyable but the main thing is, it prevents any serious medical events. There is a downside: some people do find it hard to come off the anti-withdrawal medication.

TREATMENT

After the detox, you'll be offered specialist treatment with addiction services. You may also be offered this by your GP if you are not dependent but you have had a brief intervention and it hasn't worked. In most cases, if you are considered needy enough, treatment will involve either one-to-one or group psychological therapies.

MEDICATION

There are medications to help people reduce their drinking or stop but very few people are offered them. Even though GPs see so many alcohol-related problems, sadly very few have been properly educated about what and how to prescribe.

The same is true in specialist addiction services, despite the fact that doctors there do prescribe drugs to treat addictions, for example to heroin and tobacco. It's worth raising the possibility of medication with your therapist – these are licensed medicines with proven efficacy in controlled clinical trials. They are used much more widely in the private sector.

Drugs are also much more widely prescribed in other countries. If you lived in Italy or Austria, you would likely be prescribed sodium oxybate (see below). If you lived in France, there's a very good chance you'd be put on baclofen, and if you lived in the US, you might be prescribed naltrexone. In Britain you may be prescribed nalmefene, which is used in people who are still drinking to help reduce bingeing, or acamprosate, which helps prevent relapse in people who've stopped.

1. **ACAMPROSATE (CAMPRAL)** This has been around for nearly twenty years. It's an anti-glutamate drug, which means it reduces cravings for alcohol and possibly also symptoms of withdrawal. It's been shown to work – although not in everyone – and to be cost-effective, but it's underused for reasons I don't understand. Sometimes it is used in combination with Antabuse.

2. **NALMEFENE (SELINCRO)** Approved by the European Medicines Agency to help reduce alcohol consumption in people who binge-drink, nalmefene became available on the NHS around 2014. It works on reducing the pleasure of alcohol, like naltrexone (see below). Because most GPs aren't educated in alcohol-related problems, it is only very

rarely prescribed, despite commissioning groups being told they should support it. It's widely used in France.

3. **DISULFIRAM (ANTABUSE)** This makes people vomit and flush and feel very ill with chest pain and a fast heartbeat when they drink, by blocking an enzyme needed for the body to process alcohol. People used to be prescribed Antabuse as part of aversion therapy. The idea was, after they had relapsed and experienced the results, the next time they were tempted to drink they would remember how ill they'd been. The problem is that your memories of unpleasant experiences tend to fade, and the desire for alcohol can overwhelm them. Consumption also has to be supervised as, if you don't take it, you won't get the (desired) bad effects.

4. **NALTREXONE** Designed to reduce relapse in abstinence, like nalmefene it works by blocking the endorphin system. This takes away the pleasure of drinking and so can help people who lapse avoid a full relapse. It is also used to treat opioid misuse.

5. **SODIUM OXYBATE (ALCOVER)** This is a very interesting drug, discovered in Italy by a great pharmacologist, Professor Gian Luigi Gessa. He bred a strain of rat called the Sardinian alcohol-preferring rat, then looked for drugs that would stop them preferring alcohol. He first showed that sodium oxybate stopped rats drinking as well as preventing them having withdrawal symptoms, then that the same happened in humans.[15]

Not yet licensed in the UK, sodium oxybate can be used in private clinics on a named-patient basis. It's been used in Italy for 20 years and in Austria for nearly that long and it's currently (in 2019) under review by the European Medicines Agency. There's been a movement for several years to get it licensed in Britain but it has failed. One reason is that, in its original, liquid form, sodium oxybate is used recreationally; it has the street name of GHB. So while this drug would help alcoholics, it's currently denied them because of the fear that it might be abused.

The other argument against it is that its effects are too much like those of alcohol, so someone taking it is swapping alcohol for another dependency syndrome. The best analogy here is between vaping and cigarettes. As a pragmatic harm reductionist, I say being dependent on the nicotine in vaping isn't going to harm you nearly as much as smoking cigarettes. In a similar way, being dependent on sodium oxybate is going to harm you way less than alcohol.

Sodium oxybate has a newer formulation (Alcover), a sherbet, which no longer gives users a buzz. There is good evidence that Alcover helps very heavy drinkers, people who are in the last decade of their life because of alcohol. As the only drug strong enough to compete with alcohol, it could be life-saving for so many.

6. **BACLOFEN** Originally developed in the 1960s to treat spasticity, this drug was discovered as a treatment for alcohol by a famous French-American cardiologist Dr Olivier Ameisen, an alcoholic who wanted to treat his own illness. When he started to take baclofen, which works on the GABA system, he found he was able to stop drinking.

He wrote a book, *The End Of My Addiction*, about how he cured his alcoholism. It became a best-seller in France, but there isn't as yet any good trial evidence that baclofen works. This might be because you can get tolerant to it, so people are often given too low a dose. It can be prescribed in the UK but only off licence.

PSYCHOLOGICAL TREATMENTS FOR ADDICTION
Alcoholics Anonymous (AA)

The most famous talking treatment of all, AA is effectively group therapy plus multiple other elements. The 12-step programme is a sophisticated psychological process made simple. It covers your life history of drinking, so it gets you to work through how and why you drank as well as the problems you caused for yourself and others. It helps you get insight and understanding into your reasons for drinking. It then helps you develop mental strategies that support your abstinence. The group element adds peer pressure and you also have a buddy – your sponsor – on call who has been through the whole process. Plus there's lifetime follow-up. But the greatest benefit is that you can find a meeting in every major town in the UK.

Many people in AA stay sober for the rest of their lives. The one kind of drinker who I've seen don't find it helpful are those drinking to deal with an anxiety disorder, because they find it intimidating to be in a group.

Motivational interviewing

The aim of this therapy is for people to understand the reasons for their drinking, then develop strategies to overcome them. It's essentially about replacing the pleasures and rewards of and the desires for alcohol. It helps you reframe your thinking so you can get similar or more pleasure from not drinking.

For example, if you drink to deal with work stress, you will learn new ways to reduce stress. And you'll develop strategies to help you make the tough decision not to drink after work. It will help you find new evening activities, such as sport, so you don't find yourself in the same old drinking haunts. It will also help you think through what you can do at a work function when you'd usually drink. Or, if you often drink and then argue with your partner, it will teach you to reframe this as: if I don't drink, I will have a better relationship.

Cognitive Behavioural Therapy (CBT)

The cognitive part is to help you understand the thinking processes that are behind drinking. As people often engage in behaviours without thinking, analysing them helps you gain mastery not only over them but also over their associated thoughts.

There are four elements to CBT: you work out the factors that lead to you drinking. You look at what is happening internally: is it to deal with anxiety or depression, is it because you crave alcohol? Then you look at the cognitive processes related to that; what are your expectations of drinking? Do you have a good time? Are you denying the harms? You also focus on the consequences of your drinking. For example, do you remember the fun parts and forget the fact you threw up or were arrested?

WHAT TREATMENT SHOULD I TRY?

The only thing I'm confident of clinically is that people who have any underlying psychological or psychiatric problem such as depression, anxiety, PTSD or bipolar disorder, who then become dependent on alcohol, should have these problems treated first (see Chapter Four on mental health).

After that, treatment is a question of trial and error. We don't yet know how to identify the people who will do best on any one drug or talking treatment; all we can do is give them what we have and see if it works.

A US study in the 1990s called Project MATCH, which cost $27 million, matched people with different talking therapies. The idea was to find out if that would increase the success rate.

One group had Cognitive-behavioural Coping Skills Therapy, which helps people to deal with poor self-esteem and thoughts of failure. The second group had Motivational Enhancement Therapy, which helps people to build on their strengths. The third group did a 12-step programme derived from AA. After 12 weeks, the results showed that you could not predict which programme would work; in fact they all worked equally well.[16]

FLASHPOINTS FOR OVERDRINKING

As you've read, it's easy to tip over from being a normal drinker into one who is dependent. These are the times of your life when you are most vulnerable.

● Divorce or a break-up

When people split from a partner, it's a cliché but they can end up engaging in solitary drinking to numb the emotional pain. You might have heard the urban myth about the woman who was found dead in her flat by the police breaking in and how they also found hundreds of wine bottles and hundreds of uneaten pizzas. The story goes that she'd sit at home and order a bottle of wine and a pizza to be delivered . . . but never eat the pizza.

On the other hand, because most of our socialising is done around alcohol, if the newly single want to meet people, they may feel under pressure to drink. And then there's building your Dutch courage for dating, too. But the main effect is thought to be that, when you have a partner, they look out for your health, including if you are drinking too much. A Swedish study showed that divorce leads to a significantly higher risk of being diagnosed with alcohol-related disorders, but that remarriage lowers that risk.[17]

● Going to university

Not only are young people set free from any parental shackles, but there is a drinking culture at most universities and plenty of peer pressure to join in (although not all students do). In a 2018 National Union of Students survey, three-quarters of respondents said that there was an expectation for them to drink to get drunk; nearly eight out of ten agreed drinking and getting drunk is part of university culture.[18] Tragically, this does kill young people every year: take the case of Ed Farmer, a first-year economics student who died after an initiation ceremony with the university's

Agricultural Society, which involved a bar crawl and 100 triple vodka shots.[19]

- **Death of a child or a loved one**

People use alcohol as part of their coping strategies, to deaden the pain of loss. It seems this association is highest two years after bereavement, according to a Hungarian study. At this time, the proportion of men whose levels of drinking put them at risk of health harms reached nearly 30 per cent.[20]

- **Job**

It's not unusual to drink to deal with the work stress. But the biggest risk comes when people lose their job and find themselves with days and days of free time and less reason to get up in the morning.

- **Retirement**

The number of older people dependent on alcohol or who drink heavily has risen fast. It's a consequence of an older population but also what happens when people retire and find themselves with disposable income and with more time to drink and no reason to put on the brakes. Added to that, due to other illnesses, changes in metabolism and medications, it's likely that older adults are at greater risk from alcohol-related harms.

- **Moving abroad**

Similarly to retirement, people often end up with a lot of free time on their hands. The social life of the expat may be based around drinking, and booze is often cheaper, too.

WHAT CAN YOU DO IF SOMEONE YOU LOVE DRINKS TOO MUCH?

If you want to talk to someone about their drinking, don't do it when they are drunk. It's completely pointless; people can't properly engage their brain when they are intoxicated. And chances are they won't react well.

We know that pressure plus support from family and loved ones can be a powerful way to get people to want to come off alcohol. However, a doctor can't make a spouse or partner go along to a meeting; the alcoholic has to make that decision. If you are the partner or relative of someone who drinks too much and can get your loved one to talk to you about alcohol, it's a good start.

WHAT TO DO IF YOUR PARTNER IS DRINKING TOO MUCH

A lot of people would refuse to start a relationship with someone who smokes, and I'd say it'd be sensible to do the same for someone who drinks too much. Obviously this isn't practical advice for a relationship at a later stage.

If your partner has mentioned they are worried about drinking too much, it's a good idea to monitor their drinking, the same way you might observe and comment on what he or she is eating if they had told you they wanted to cut back. That way, you can be the nudge for them to slow down or stop. By doing this, you can hopefully help them before they become alcoholic.

However, I'm assuming that if you're reading this, you are already past this point.

If you are worried, the first thing to do would be to note what they are drinking over a few weeks, and any adverse consequences. And at some point – when they are not drunk, you have the time and you are alone with them – have a conversation about it.

If your partner says they understand, together you can develop a strategy to reduce the harms of their drinking (see Chapter Nine for details of how you might do this). It might be that you agree set nights to drink or a set limit of drinks per night.

If they need convincing, it's worth giving them evidence of their altered behaviour; for example, you could film them. You can talk through the problems that their drinking is causing, whether it is destroying your relationship, or their work, or their relationship with any children.

Rather than get angry, try to be calm and clear in your analysis of their problem. There's a real opportunity, during the contrite phase after drunkenness, to get them to see that controlling their drinking is not only in your interest, but in theirs too. Perhaps you could ask them to read this book, for more arguments on why it's a good idea to stop or cut down their drinking?

You could also ask them to make an appointment with their GP for a health check-up. GPs are trained to ask about and pick up on harmful drinking, and a diagnosis of high blood pressure or an abnormal result of a liver test or scan can be a powerful motivator. If they have an underlying psychological issue that is leading them to drink – such as bipolar disorder, depression, anxiety or OCD – encourage them to have treatment for that.

ADDICTION: HAVE I GOT AN ALCOHOL PROBLEM?

If the issue is stress at work, you can suggest alternative stress-reducing behaviours.

If it's more your partner's social behaviour when drunk, counselling might be a good idea, either together or for them alone, so you can have an independent person helping you work through the issues.

You don't have to leave someone just because they are an alcoholic. After all, it is a mental illness, like depression, and most people wouldn't leave someone just because of that. Many people can and do overcome their addiction. But if your partner is violent or refuses to seek treatment and it is affecting your or your children's lives, it may be time to leave. As well as your own GP, another place you might want to try is Al-Anon, a group with a 12-step programme that exists to support people through the problems of being in a relationship with an alcoholic.

WHAT TO DO IF YOUR CHILD IS DRINKING TOO MUCH

This is challenging because you won't know how much your child is drinking if they are drinking socially, separately from you. But they are probably drinking too much if they are coming home drunk and vomiting, and/or having terrible hangovers. Especially if they are doing this regularly.

Often an issue for teenagers or young adults is that they'll say they are only doing what their mates do. It's your job to explain that doesn't make it safe. You need to base the discussion – assuming they will have one with you – on reason and evidence. You could use some of the facts in this book or ask them to read it themselves.

You could also start by asking them: might you be drinking more than your friends? How do you feel after you've been out drinking?

Explore ways that life might be better if they didn't go out and get drunk. Could they do more of a sport they love? Or another activity? Could you do something together with them in your free time?

If your child refuses to engage in conversation, it is very hard. Can you get support? There are expert charities for this (see the Resources section at the back of this book). At this point, if you are giving them money and they are spending it on drink, stop. Also let their GP know, if they are below the age of consent.

WHAT TO DO IF YOUR PARENT IS DRINKING TOO MUCH

If you are a child in this situation, please speak to someone. Can you talk to a teacher? Or a relative? You could call Childline or Nacoa, a charity for people affected by their parents' drinking (see page 265). You are not the only one: we know roughly one in ten adults are alcohol-dependent, so that must mean that about one in ten parents are too. We know there are schoolchildren who have to look after their drunk parents. And we know that, for some children, that is what happens every day.

When I took part in a *You and Yours* phone-in programme on Radio 4, the topic of how other people's drinking had affected the callers' lives got their largest ever number of callers at the time. Many were older people who were still distraught and distressed by what had happened to them at the hands of a parent many decades before.

If you are an adult with a parent who is drinking too much, see the advice given above for partners. You can also seek help from the charities mentioned on the previous page.

WHAT TO DO IF YOUR FRIEND OR RELATIVE IS DRINKING TOO MUCH

All the advice given in the section on partners applies here, too. This can be particularly hard because you may not be living with that person, and they may be less willing to speak to you. In fact, you may find that they resist talking to you and deny that drink is affecting them. They may also turn against you and end your friendship or contact; but at least you'll have done the right thing. And if in the future they reform their behaviour they will likely return to thank you for helping initiate this change.

See the Resources section at the back of the book for charities that can help.

COMMON QUESTIONS

SHOULD I WORRY IF I'M ALWAYS THE LAST ONE LEFT AT THE PARTY?

Some people don't seem to know when to stop. And it may be their brains are different from the norm. For them, the endorphin rush becomes so pleasurable that they lose the ability to make the decision to stop drinking, and just keep seeking the rush. Even though they may start off the evening wanting to drink normally, the alcohol changes how their brains work.

The genetics behind being more likely to binge-drink isn't known. What we do know is that if you are a binge-drinker who loses control, you can bring yourself back towards normality by blocking the endorphin system. The drug nalmefene may help, as it works to stop the effect of the endorphin release. Try controlling or stopping your drinking (see Chapter Nine) and if this doesn't work, see your GP.

I STILL GET UP IN THE MORNING AND GO TO WORK, SO HOW CAN I BE AN ALCOHOLIC?

A critical part of a diagnosis for any mental disorder is impairment. Alcohol affects your judgement; you can be impaired and not know it, so you might not be the best judge of whether you're an alcoholic. I'm afraid that going to work simply means you go to work.

Ask your partner, a trusted friend or family member to tell you the truth about your drinking. This will give you your best chance of being confronted with the reality of your behaviour.

WHY WON'T MY PARTNER EVEN DISCUSS THEIR DRINKING PROBLEM, LET ALONE TALK ABOUT STOPPING?

It's common for people to be in denial about a drinking problem. They don't want to believe that alcohol is an issue, because they need it. Most people think their judgement when under the influence of alcohol is OK, or even that it's better than when they're sober.

There are very few people who consciously want to be an alcoholic and destroy their lives. After all, everyone can stop drinking; you just don't pick up the bottle. The reality is that people who are alcohol-dependent can't stop because their brain won't let them stop. When someone is at that point, it's not an intellectual decision. Their brain is beyond reason, has modified so alcohol has become the driving force of their life. People who arc physically dependent can even have drinking dreams, as well as thinking about alcohol all the time they're awake.

There is a tension that exists with drug use; that the pleasures outweigh the harms. The most common reason behind someone ending up in treatment is that they've lost something: their job, partner, kids or their liberty. People often only make the decision that they want to change when their life reaches crisis point and they finally realise that the alcohol has contributed to or caused it.

Some people will still refuse treatment. One extreme example of this was a 50-year-old Roman Catholic priest, an intelligent man and a university graduate.

When I met him he was living in a Salvation Army hostel, having dropped out of the priesthood ten years previously due to his drinking. It was very sad to hear how this man's life had slowly been destroyed by alcohol.

The priest's drinking behaviour was not untypical of an end-stage alcoholic. He got a weekly cheque every Thursday, for around £150. He'd cash it and go to the pub. He'd drink half of it on Thursday night, the other half on Friday. Then, for the whole of Saturday and Sunday, he'd be in withdrawal.

As we talked, we agreed this was a terrible way to live. I told him about the proven medications and treatments for alcohol. He said, that's very kind of you but I don't think so. His love of alcohol was greater than his love of life or even of God.

The pleasure he got from two days of drinking a week outweighed every other pleasure in his life.

His story is extreme but what it illustrates is, firstly, the person has to want to change. And secondly, in order to overcome the addiction to alcohol, they've got to put a lot of good reasons not to drink into their brain, as well as develop the skills and strategies to overcome the subconscious drive to drink. This will take expert help and possibly medication. Your partner won't stop drinking because, at the moment, they can't.

THE SOCIAL BENEFITS
OF ALCOHOL

HUMAN BEINGS ARE social creatures. And you could say alcohol is THE social drug. Of all the recreational substances we take, perhaps with the exception of ecstasy, alcohol creates the greatest sense of sociability. You could say alcohol provides the perfect drug cocktail for socialising: you become energetic and positive as if you're on cocaine, you become relaxed as if you're on GHB, and you love other people as if you're on ecstasy.

There has been a lot of work done to analyse the harms of alcohol, for obvious reasons, but there hasn't been nearly so much scrutiny of its benefits. These are both social, such as relaxing, bonding and spending time with other people, and creative, allowing us to expand our minds into new thoughts and territories.

These kinds of benefits are hard to measure. But I'd say they had and still have an enormous value in society. In fact, human beings have been making alcohol and drinking it together for thousands of years. It's the easiest drug to make – all you need is rotting fruit.

There's an anthropological theory that alcohol was a crucial element in the development of organised societies, in our species making the switch from nomadic hunter-gathering to farming. It says that crop cultivation may have begun in order to grow wheat to make alcohol as well as for food.[1]

It's also been suggested that alcohol, as a driving force behind human creativity, has fuelled the development of cultural expression, from art to music to religion.

And it's true there's increasing evidence that every society in history had a way of making alcohol, and that it was used at significant events like rituals and for special occasions. It's only in the past millennium that it has been banned by some religious groups, for example in Islam.

Honey, for example, was fermented into a strong, sweet drink called mead. Some experts think this may even have pre-dated the fermenting of wheat, that people were drinking mead in Africa over 40,000 years ago.[2] It's thought mead became popular in northern Europe because it was harder to grow wheat there. In the early Middle Ages, it reputedly fuelled Viking bonding ceremonies and warfare (see Chapter Three) a bit like it does for some rugby clubs today.

Distillation of wine into spirits came much later, probably in Arabia about AD 900. In fact, the word alcohol is derived from the Arabic *al-kuhl*, meaning 'body-eating spirit' (for some people, this is sadly true). The distillation technology reached Europe, where in the thirteenth century Arnaldus de Villanova, a physician, commented: 'We call it *aqua vitae* [life water], and this name is remarkably suitable, since it is really a water of immortality. It prolongs life, clears away ill-humors, revives the heart, and maintains youth.'[3] We know more now, of course.

But, for most of us, alcohol does have a broadly pleasurable effect.

A group of French researchers set out to measure the benefits and harms of alcohol compared to other addictive substances, using a similar method of analysis to those done on harms only by my group.[4]

They asked addiction experts to put a value on the benefits we get from the most commonly used addictive substances and our behaviour (including alcohol, tobacco, cannabis, cocaine, heroin, amphetamines, ecstasy and gambling) in order of their impact.

The researchers broke the benefits down into six categories.

1. Hedonistic benefits. The intensity of the pleasure obtained.
2. Identity benefits. How it allows an individual to socialise and be part of a group, plus its cultural value.
3. Auto-therapeutic benefits. How it soothes internal suffering and tension.
4. Economic benefits. The value of production, sales, consumption etc to society.
5. Social benefits. How it helps maintain social balance.
6. Cultural benefits. How it's used for celebrations and social rituals.

As in my group's analysis, alcohol came out as the most harmful addictive substance to society. But it was also the one perceived by the French researchers to have the most benefits. In fact, it was so much better than the other drugs and activities, they called it an outlier.

My question to you is, what benefits does alcohol bring to your life? Nearly all of our celebration rituals revolve around alcohol, from the cradle to the grave. We wet the baby's head, we toast exam results and birthdays, we crack open the fizz at both wedding breakfasts and divorce parties. And we say goodbye to our loved ones with alcohol at a funeral wake (and drown our sorrows too). When you are working out how much to drink – or not drink – you may want to factor in the social benefits you get from it.

DRINKING: THE GOOD TIMES, OR WHY WE DRINK

The most common reason people give for drinking is having a good time with others. How much of a good time was measured in a study that used an app called Mappiness. The app asks you to measure how happy you are on a scale of zero to 100, then to plug in where you are, what you're doing and who you're doing it with. While they were drinking, people reported being an average of four points happier.

Another app-based survey by Pebble (a now discontinued smartwatch) confirms what we know about the feel-good glow of drinking. It asked people for their self-rated level of happiness as well as what they'd done in the previous hour, and drinking came out as giving the best buzz – followed by yoga, exercise, socialising, mediation, food, then caffeine. It scored pretty highly as being energising, too.[5]

People also use drink to change their mood, to reduce negative emotions and provoke positive ones, at least in the short term. This isn't surprising, considering alcohol's endorphin and dopamine brain hit.

Alcohol lubricates our interactions with others too. If you feel nervous about talking to people at a party, as soon as you have your first drink, it helps you overcome that threshold of anxiety so you can talk to people. A glass to hold gives us something to do with our hands in awkward situations, particularly now smoking has become so vilified. Go out for dinner and you add even more of the good things, like socialising and food, into the mix, too.

There's another level to this. Research shows that people who drink socially – especially those who have a 'local' – tend to have more friends and so more emotional support, a key source of mental wealth. They also feel more contented and more involved in their local community.[6]

Alcohol is a major part of how we bond. Asking someone to go to the pub is a culturally acceptable way to make friends, especially for men. Half of the reason why people spend their evenings or weekends playing in sports teams is probably the bonding in the bar after the game or match.

Drinking has also been shown to promote social feedback: facial expressions of interest, for example, as well as verbal ones. So you look more animated, more interested and you feel as if you're being more interesting (hopefully you're not the boring drunk, droning on!).

In a study, when strangers were put together in groups of three, those given alcohol were more likely to smile fully, and smile at the same time as each other. The drinking groups were also more likely to have everyone joining in the discussion.[7]

Drinking may make you less aware of other people's criticism. But it may also make you feel that people are more positively disposed towards you, at the same time as you become more positively disposed to others too. Good feelings all round![8,9]

Research suggests that drinking in groups – in particular binge-drinking – takes the positive feelings up a level, to euphoria. A prime example of this are the stag or hen weekends in European cities where the booze is cheap (although it's best not to start drinking before getting on the plane as you may be kicked off, along with the rest of your group!).[10]

THERE IS A DOWNSIDE

While people like to drink together, it's easy to drink too much. One study found that there were four main motives for drinking more in a group: copying (automatically copying another's drinking behaviour and how much they drink), conformity (fitting in with the group), hedonism (escape and having a good time) and winding down (relaxation). And once we've used alcohol as a stress valve, that behaviour becomes embedded.[11]

IS YOUR BOOZE BUZZ IN YOUR HEAD?

Something that might make you think – and especially for anyone who wants to stop drinking – is that you can get the buzz without the booze. In a study in New Zealand, people were put in a room that looked like a bar. Half of them were told they were drinking vodka and tonic, half just tonic. But, in fact they'd all been given tonic. The fake drink glasses were rimmed with limes dunked in vodka so they smelt of alcohol, but the alcohol content was negligible. The – totally sober – people acted drunk by flirting and giggling and, even when told otherwise, believed they had been drinking.[12]

For many people, in many circumstances, alcohol does what you want or need it to do. If you want to be happy and have fun, it can do that. If you want to numb your feelings, it can do that. If you want to bond with the group and let loose, it can do that too.[13,14]

DOES ALCOHOL CHANGE YOUR PERSONALITY?

How does your behaviour change when you drink? Possibly less than you think. In a lab study, groups of three or four people were given alcoholic or non-alcoholic drinks, then asked to complete puzzles or discuss questions together. The tasks were designed to help reveal their personality traits.

When people were asked to rate themselves, they thought that after alcohol they became less self-disciplined, less open to trying new things, less agreeable, but more emotionally stable and extrovert.

But when the rest of the group rated them, the only change the others noticed was a boost in extrovert traits – that they were more gregarious, active and assertive. The conclusion? Alcohol doesn't change you as much as you think.[15]

If people do transform under the influence – and we've all seen people who cry or get angry when drunk – it may be because alcohol's disinhibiting effect is allowing any repressed emotions to come to the surface. Most of the time we try to control our behaviour, to keep our emotions in check in a way that feels socially appropriate. When you drink, you can just let it all hang out.

So if you're miserable at work but putting on a brave face, if you drink you may end up upset, 'drowning your sorrows' or at least sharing them with a friendly and receptive audience!

Or alcohol might dissolve your facade and liberate your underlying aggression. According to research, the personality trait that predicts this is if you tend to live in the moment rather than consider the results of your actions. This finding held true for both men and women – although men were more likely to be aggressive.[16]

THINK: WHY DO I WANT TO DRINK?

Which one (or more) of the following motives describes why you drink? It will help you decide which drinks are worth having or if another behaviour might work instead.

THE SOCIAL BENEFITS OF ALCOHOL

1. You want to get off your head because it's exciting and different. You no longer want to be in control of your life.

2. Peer pressure. This might be buying rounds or drinking in a group. In a more extreme form, it's drinking games. In 2014, it was reported, up to five people died in the UK and Ireland doing the neck nomination or neknomination drinking game, an online dare to chug that you then had to pass on to another person. Of course, drinking games aren't new: the old-fashioned version was to down a yard of ale.

3. You want an escape from having to think or feel, an emotional anaesthetic that takes you out of the drudgery of life. After two to five units, as the endorphins and serotonin kick in, you no longer care so much about your problems as your brain no longer functions that way.

4. For the laugh-out-louds and the highs. Dopamine and endorphins are what gives you the belly laughs and excitement so you don't know where the night will lead. You feel animated and interesting.

5. You want to have sex and having a drink is how you feel brave enough to meet and/or talk to potential dates.

DRINKING AND DATING

There's a classic phrase, already mentioned, to describe how your perception of people distorts under the influence of alcohol: beer goggles. That's when a few drinks turns the man, woman or person you don't fancy into someone who's extremely attractive.

DRINK?

In research that asked people about first dates, women said that drinking increased feelings of intimacy – defined as closeness, affection and connectedness. For men, the opposite was true.[17]

In one study, people felt more positive towards pictures of others after a drink. But they didn't think art was better when they were drunk, which suggests it's a social effect rather than an aesthetic judgement.[18] And a second study showed that people find tipsy faces more attractive than sober ones – but drunk ones less so.[19]

So is drinking and dating a good thing? Not necessarily. In another study, 62 men in their twenties were sent on virtual dates with a computer simulation of a woman. Those who drank were much more persistent sexually in their conversations.

Probably in the majority of cases, alcohol lubricates what can be a tense and anxious situation, especially if it's a first date. At least there will be more relaxed conversation and increased good humour. And if one of you drinks more than is acceptable, then at least the other person has the option of turning down a second date.

HOW TO DRINK THE WAY YOU WANT TO (AND SENSIBLY)

SO YOU'VE DECIDED you want to drink less. As you know by now, I'm not a prohibitionist. I want you to get the maximum benefit out of drinking, with the least amount of harm.

The issue is, alcohol is moreish. If you like to drink, it's always going to be easy to drink too much. In our alcohol-centred culture, if you say yes to every pint, glass or bottle, you will end up drinking too much.

So it's up to you to discover the strategies that help you drink to the limit you decide, that suit your finances, time limits and lifestyle.

Ideally, if you find those that work, you'll be able to find an amount to drink that you are satisfied with, where you can function, have a good time and feel comfortable with your level of health risk.

The first step is to think about what kind of drinker you are because the reasons you drink will make a big difference to the strategies that work. Next, I suggest a whole range of strategies, so you can pick and choose.

In this chapter, I'm not talking to dependent drinkers. If you are dependent, you will need medical attention during withdrawal and, most likely, a more structured approach to cutting down or giving up. Although some of these strategies may be part of that. For more, see Chapter Seven.

WHAT KIND OF DRINKER ARE YOU?

You may think: why do I need to know this? Without analysing our drinking, we end up drinking too much. There are just too many opportunities to have alcohol. And, as I've said before, it's a strong drug. You wouldn't take medication or vitamins or an illegal drug without thinking about it first.

Most of the data over the last century has suggested there are two patterns of people who become alcoholic. These may not seem directly relevant to you, but they are enlightening about how alcohol can take hold of people.

The first group are young men with a family history of alcoholism. They become dependent almost from their first exposure to alcohol: with their first drink, everything makes sense.

The second group are people who drink to deal with the problems in their lives, whether those are anxiety, depression or stress. They tend to turn to alcohol for self-medication from their thirties onwards.

In the normal pattern of drinking over a person's lifetime, there are life phases when you are more vulnerable to drinking too much.

Young people start out drinking socially for pleasure, and often due to peer pressure too. They want to relax and they like the feeling of losing control.

Some people never stop drinking like this. But most give it up when life gets more responsible, whether that's due to a job or a relationship or a family.

Often, people will resume drinking later in life to deal with stress. And those with bigger issues, such as mental illness, grief or PTSD, are particularly vulnerable in their search to suppress their symptoms.

One interesting way to classify your drinking comes from Professor Emmanuel Kuntsche and Dr Sarah Callinan at the Centre for Alcohol Policy Research, La Trobe University, Australia.[1] It's useful because it helps you analyse your motivation to drink, rather than all the specific one-off reasons you might have, such as a birthday or a hen night. The idea behind this analysis is that your primary reason to drink is to change the way you feel, either to make more of positive feelings or to dampen down negative ones. Every other reason is secondary.

In this model, there are four types of drinkers. The categories are based on young adults, but in my experience, there are individuals of all ages who are typical of each of these categories and most of us have some kind of overlap.

1. **Social:** You drink mainly to celebrate and have fun.

   Are you preloading, drinking before you go out? Do you have a good time when you don't drink? Can you imagine a social occasion without drinking or do you feel you need to have a drink in your hand? Social gatherings are difficult at any age, and social drinking can become a way to cope with them.

2. **Conformity:** drinking to fit in. You drink because other people do. If what your friends do is go out drinking, it's very difficult to be the one who doesn't drink.

   Are you drinking in rounds or doing shots, feeling you have to keep up with other drinkers? Does it feel rude to refuse a drink or someone topping up your glass? Would you be OK to be the person at the party who doesn't drink?

3. **Enhancement:** drinking because it's exciting. You might be a risk-taker or do things when drunk that you wouldn't normally do. You're likely to go out with the intention of getting drunk.

   Do you crave the feeling of being drunk? Do you need to be drunk to let yourself go? Do you tend to binge-drink? Don't wait until you have a bad experience or get injured or need a health intervention before you deal with your drinking behaviour. If you do drink in order to get drunk, start to take cutting down very seriously.

4. **Coping:** drinking to forget your worries. This is the largest group, and in my experience includes a lot of middle-aged people who drink alone, self-medicating. These are the people who are most likely to become alcoholic. If you are drinking for stress reduction, do seek help rather than

papering over the cracks – or, rather, filling them up with vodka.

Do you drink every day? Do you use alcohol to numb your bad feelings? Are you drinking more than you used to? Are you able to spend a night alone without drinking? Start to think about ways you can learn to cope without alcohol.

PERSONAL HEALTH STRATEGIES FOR ALCOHOL: WHICH ONE WORKS?

First you need to really get a handle on your drinking: how much and when, as well as why and with whom.

WHAT IS A UNIT?

People who've never counted units are often shocked to discover how much they're drinking.

In the UK, one unit is 10ml or 8g of pure alcohol. To give you a rule of thumb, what follows are the NHS guidelines.

- 2.1 units in a standard glass (175ml) of average-strength wine (12 per cent)
- 3 units in a large glass (250ml) of average-strength wine (12 per cent)
- 2 units in a pint of low-strength lager, beer or cider (3.6 per cent)
- 3 units in a pint of higher-strength lager, beer or cider (5.2 per cent)
- 1 unit in a single measure of spirits (25ml)

However, the strengths of wine and beer, and glass and serving sizes, vary widely. So I suggest that, both in the places you go out to drink and at home, you work out your average units in advance. Here's how to do it:

How to calculate your beer or cider units:

1) Look at the Alcohol By Volume (ABV). This is the percentage (%) on the label.

2) Work out the bottle/glass size.

If you are drinking beer or cider that comes in a measured bottle, it's easy. In a pub, it's 284ml for a half pint, 568ml for a pint.

3) Do the sum.

The sum looks like this: ABV% x ml ÷ 1,000 = units

So for a pint of 5 per cent beer (568ml), the sum looks like this:

5 x 568 ÷ 1,000 = 2.8 units

For a lower-alcohol beer, say of 2.8 per cent, and a half pint (284ml), it will be:

2.8 x 284 ÷ 1000 = 0.8 units

How to work out your wine units:

1) Look at the Alcohol By Volume (ABV).

This is the percentage (%) on the label.

2) Work out your glass size.

In pubs, small is 125ml, medium is 175ml and large is 250ml. By law, there should be a price displayed for a 125ml glass. But be aware: sometimes you're offered 175ml as the smallest glass.

At home, to find out the size you usually pour yourself by filling up your glass to the usual level with water, then pour it into a measuring jug.

3) Do the sum

ABV% x ml ÷ 1,000 = units

So a 13.5 per cent wine in a 175ml glass, the sum looks like this:

13.5 x 175 ÷ 1000 = 2.4 units.

How to work out your spirit units:

Pub spirits are often 25ml – which is one unit – but some bars serve 35ml or even a double measure, 50ml, as standard. Spirits are usually 35–40% ABV; however, there are some that are stronger, around 50%.

IS CUTTING DOWN RIGHT FOR ME?
OR SHOULD I GIVE UP?

If some of the below ring true for you, you might want to consider giving up rather than cutting down.

What follows are indications that cutting down might not be right for you:

1) You have tried and failed to give up before. You may want to seek help from your GP because you may be dependent. The same is true if you drink heavily (see addiction advice, Chapter Seven).

2) Your medical history.

DRINK?

All of the below conditions/habits/health risks have been linked to drinking and/or made worse by drinking. If any apply to you, you may be better off giving up.

- You have a mental health condition such as depression or anxiety.
- You suffer from gastric upsets: indigestion, gastritis or reflux.
- You've given up smoking – or want to. (A lot of people relapse back to smoking when they drink, finding it to be a trigger.)
- You have high blood pressure.
- You have diabetes or prediabetes.
- You are at risk of stroke or heart attack, or you've had either.
- You have a family history of the cancers linked to alcohol (see page 54).
- You have epilepsy. (You're more likely to have a fit when in withdrawal from alcohol.)
- You have a history of other drug use. This tends to go up when people drink.
- You want to conceive. This goes for both men and women.
- You have insomnia.
- You want to control your weight. Alcohol contains significant calories (see page 124).

3) Your personality

If your issue is that you lose control around alcohol, you may find it's better to take an active decision not to drink at all. It can be easier to say 'I don't drink', as opposed to,

'I'm not drinking tonight,' which can provoke the question, 'Why, don't you like our company?' Peer pressure to drink can be quite powerful and I often hear stories of people drinking when they didn't intend to.

If you go out intending to drink a little, you run the risk of failing. But if you decide not to drink and stick to that, people might think you're a little unusual but at least you've succeeded. And in fact, the good news is that being teetotal is becoming more socially acceptable, especially in younger age groups.

How to keep a drink diary

I've found keeping a daily drink diary to be the most useful way for people to get a handle on their drinking.

WEEK 1

● Fill out your drink diary

Include the time, where you were, who you were with, what you drank, how you were feeling and why. At the end of a week, for extra motivation, fill in a survey like the one at drinkmeter.com. It will compare your drinking with other people's and tell you whether or not you're drinking at dangerous levels.

● Write down why you're cutting down

Stick this up where you'll see it often, on the fridge for example. Your reasons could be health-related: I want to be healthier for my children, I want to have more energy, I want to feel more positive, I want to sleep better, I want to give up smoking. Or they could be linked to better performance: I want to run a 5k, I want to get a promotion at work. Or they could even be linked to looks: I want my skin to look more glowing, I want to lose weight, I want to age more slowly.

WEEK 2

• Eliminate the drinks that weren't worth having last week
Looking at your week of drinking, rank your drinks in order of fun, satisfaction and results.

Which ones left you feeling good? Which could you have done without? Which do you wish you'd refused?

• Write down your triggers
On the nights you drank too much, what were you doing? Did you come straight out of work and, stressed, go to the pub to complain about it? Did you go home and pour yourself a large glass of red? Or was it that you were sitting alone and watching TV, and fancied a glass? Or that you were in the pub and everyone was buying rounds? Can you avoid these triggers? Work out what you'll do next time you're in that situation.

• Write down how you drink
Drink faster and you'll get drunk faster. Do you drink fast or slow? Does it depend on what you drink? Can you find a drink you can drink slowly? For example, you might find you drink red wine much more slowly than white. Can you slow down by alternating with soft drinks?

WEEK 3

• Set your target threshold

Decide how many units you're going to have next week. Calculate how many of your preferred drinks that is.

This has to be a threshold that you know you can keep to. If you set a threshold of a maximum of four units any one evening but you know you always lose control at four, then you should go lower. Remember, alcohol is one of the worst drugs for judgement.

• Be very specific

Look at your nights out and/or other possible drinking opportunities for the week and work out what you're going to drink on each occasion to avoid the 'oh just one more' trap. Work out if you're going to drink beer or wine or spirits – then work out exactly what drinks you'll have (use the unit calculator on page 194). How are you going to spread out your drinks over the week – bearing in mind you should have at least two drink-free days a week.

• Decide a way to monitor your drinking

Some people use tokens to count units, swapping them from one pocket to another. Some people write it in a notebook or on a card they keep in their bag. Some people use apps, for example Drinkaware. However, the truth is, if you're drinking sensibly you're very likely to have four units – usually two drinks – at the most, so you should be able to count them! However, many people do find it satisfying to write down their units.

If you go above your self-set limit, think about why this has happened, to help you next week.

WEEK 4

• Change your routine

If you want to, which drinks can you cut out this week? Are there any that you're drinking just out of habit or which are no longer fun? Are you drinking out of stress? Adopt five of the suggestions in this chapter that make sense to you.

From now on, try out different strategies each week, until you reach the level of drinking that you like.

• Decide what your reward will be

Every time you keep within the limits you've set yourself, decide what your prize will be (tip: not a large glass of Merlot).

SENSIBLE DRINKING RULES YOU MIGHT WANT TO TRY

You don't need to do every one of these. In fact, there is some crossover between them, so you can't.

The idea is to try out a few you think might work and see if you want to – and can – adopt them permanently. I can't guarantee they will all work; some are evidence-based, some are personal rules that patients and colleagues have told me, and some I do myself.

What you are aiming for is a few new and lasting habits that help you keep to the limit you are happy with.

HOW TO DRINK THE WAY YOU WANT TO

Don't drink when you're thirsty

Always satiate your thirst first with a soft drink or just water before you start on your usual full-strength drink. If you are a beer or lager drinker you might even try a shandy. Some people find that if they drink beer first, they drink less over the evening because it fills them up – but bear in mind it's very easy to go over your limit when you are drinking beer, especially strong ales and lagers.

Never allow anyone to top up your glass

At parties and dinners where the wine is flowing, you're bound to be offered more before you've finished your glass. Make it a rule that you must empty the glass before it's refilled, so you can know exactly how many glasses you've had.

Say no to champagne

The same goes for prosecco or any other sparkling wine. You're often offered it as your first drink on an empty stomach, the bubbles mean you get intoxicated faster, and at 12 per cent ABV, it's relatively strong alcohol. This kind of drink might be part of social culture – but you can say no. Oh, and if you're offered champagne with pudding, you can say no to that, too. After all, it's likely you'll have drunk enough by that point.

Don't wash down food with wine

Have a jug of water and glasses on the table or a glass beside you when you're eating. Some people find they drink more sparkling water, or water with ice or a slice added; whatever makes it more appealing. Or you could make it a rule that you only drink wine between courses, not before the meal comes and not when the food is in front of you. Really savour the taste of each (small) mouthful.

Refuse aperitifs

There is an idea that having an aperitif makes the meal more pleasurable, that it's a necessary part of eating, but it's really just a way to get drunk fast.

Only have two drinks

Two drinks will make it very likely you'll stay within the recommended limits, if you are drinking pints or medium glasses of wine. You may also decide that your limit is three.

Aim never to get drunk

If you feel drunk, it means you have probably consumed at least five units, maybe much more if you are a heavier drinker with more tolerance. And be aware that although you may not feel drunk compared with the people around you, whether that's true depends on who you're drinking with![2]

Never have one for the road

What an amazing concept 'one for the road' is! You're ready to go home and you're likely to be tired. You've probably drunk enough.

Volunteer to drive

That way, you can legitimately refuse all alcohol without feeling you might offend. However, never drive anywhere you know you might be tempted to drink. Go by public transport if you can. It's far better to drink and be safe – even if you drink a little too much – than to drink and drive.

Take a week off drinking

This isn't as effective a reset as taking a month off (see 'Dry January' on page 214). But it will help you reset your tolerance so you'll need less booze to get the same effect – more bang for your buck, essentially. One thing: do not go straight back to drinking at the level you were before. Your tolerance will be lower, so you will get drunk faster. And if you have been a heavy drinker, it can be very dangerous.

Only buy booze as needed

Some people find it's best not to keep alcohol in the house. In any case, don't 'build a cellar' to be there to tempt you. Instead, buy on the days you decide to drink.

Adopt a curfew

The traditional naval saying 'The sun is over the yardarm' was supposed to stop people drinking before noon! But you can create a healthier version with your own timings – for example, that you don't start drinking until 7 p.m. and you stop by 9.30 p.m.

Alternate your drinks

This is a great strategy if you're drinking, say, vodka and Coke: skip the vodka every other round. For beer and wine, drink a full glass of water between drinks. Spritzers are another way of diluting your drinks, although they give you false confidence: in pubs, remember that a spritzer is a full glass of wine plus water.

DRINK?

Don't get alcohol delivered

Don't use a delivery service like Deliveroo or join a wine club, especially if it delivers you a selection automatically. They anonymise your relationship with the purchase. Instead, make yourself go to the shop so you can think about what you are choosing, why you are buying it and when you're going to drink it.

Buy expensive alcohol

Or at least the most expensive you can. Thinking about how much money you're pouring down your throat may be a deterrent to drinking more than you want to. And you will hopefully also enjoy the flavour more.

Find low- and no-alcohol drinks you like

There are so many more on the market than there used to be – from pseudo-spirits you have with mixers to soft drinks that aren't as sweet or childish. The bad news is, some of them are as expensive as alcohol, some even more.

More pubs are now serving low-alcohol beer on tap and if you want to hide the fact you're not drinking, a pint has to be the best way to do it. Most pubs now sell bottles of very low (0.5 per cent) beer and some sell cider too. In terms of units, these are practically non-alcoholic. Swapping one 275ml bottle of Beck's for one of Beck's Blue will save you 1.3 units. Swapping a 500ml bottle of Sheppy's Cider for their 0.05 per cent version will save you 2.7 units. For a good range, see zeroholic.co.uk.

HOW TO DRINK THE WAY YOU WANT TO

Change your patterns

Try one of the new ways to drink – or not drink. Some people call themselves 80–20 drinkers, so they drink two days out of ten. Or you could be a 5–2 drinker, only drinking two nights out of seven. Or a Weekend Warrior, who only drinks Friday, Saturday and/or Sunday. NB: don't binge: remember the data suggests bingeing is more dangerous in terms of brain effects and safety risks, among other health harms.

Find new activities

Sport clubs are the most obvious swap for the pub or drinking at home, as exercise and alcohol don't mix. Plus they're good if you're craving the friendly buzz of the pub. But any hobby or evening class will do. Been meaning to learn upholstery? Or do life drawing? Book in! At weekends, meet friends for walks or for shopping or cups of tea instead of at a pub.

Book in drink-free days

The guidelines say two drink-free days a week is the minimum number to aim for. But even one is much better than none. Two consecutive nights will give your liver time to recover. As before, on the nights you do drink, don't binge. Use the NHS app Drink Free Days to help you.

Don't pre-drink

Because alcohol is expensive when you're out, you may be tempted to get tanked up before you go. But it's too easy to have all your units before you've even left the house.

DRINK?

Don't drink at home

Because alcohol is so much cheaper to buy from supermarkets and off-licences than it ever has been, as a nation we are drinking more than ever at home. Some people even have special beer and wine fridges! But if the best part of drinking is being sociable, why not stick to only drinking when you're out?

Don't drink alone

Drinking alone is a habit you can do without; there's nobody to monitor you or make you rethink. And especially, don't drink in front of the TV or a screen, because when you do that, it becomes mindless. You can finish a bottle of wine before you've even realised.

Don't buy alcohol at the supermarket

Whether you buy as part of your regular shop or whether you add a few bottles to your supermarket delivery, stop. Instead, make alcohol a special shopping trip. You will probably buy less and you will realise exactly how much you are spending, too.

Avoid cheap alcohol

That means happy hours, which make you drink more in a shorter amount of time, and also supermarket offers. Bargains are difficult to resist but if you take the alcohol home, you're unlikely not to drink it.

Make sure you're hydrated

Carry a refillable water bottle and keep drinking all day. The more hydrated you are, the less likely you are to down your first drink fast.

Enjoy what you drink

Choose alcohol you love the taste of, then savour it. Think of it as a delicious treat, not something to throw down your neck. Sip, don't gulp.

Never buy a wine box

They may seem like a good idea – after all, you won't feel you need to finish the bottle. But if you have one in your fridge or cupboard all the time, it makes drinking far too accessible.

Never open a second bottle

This is another one for the wine drinkers. If two of you share one 13 per cent bottle of wine you'll automatically limit yourself to five units each.

Don't eat salty snacks

Salty snacks are put out in bars specifically to make you drink more. So avoid them at home, too.

Treat your stress first

Plan your personal list of activities you can do to reduce your stress levels before you turn to drink – for example exercise, meditation, yoga, breathing exercises, call a friend, have a bath, go for a walk.

Count before you go out

Say to yourself (or to your partner or a friend, to help you stick to it), 'Tonight, I'm only going to drink two glasses of wine.' And, if you know where you'll be at each stage of an evening out, work out exactly where and when you'll have those two glasses. So you might plan your night like this: orange juice and soda at the pub, one glass of wine when you arrive at the restaurant, one glass with the main course.

Practise saying 'no thank you'

Don't be afraid to say 'no, thank you' to your host. It might help to practise; people can be very persistent. If you feel you need an excuse other than you're trying to drink less, think one up in advance. You can say you're driving, you're driving the morning after, you've got a big day the next day, you had a big night the previous night, you're on doctor's orders, you're on antibiotics or medication. Say whatever you feel comfortable with.

Order the smallest measure

If glasses were smaller, we'd drink less. A study from the University of Liverpool took place over four pub quiz nights. On two nights, the pub sold drinks in standard sizes: pints and wine in 175ml glasses. On the other two nights, they sold beer in 2/3 pints and wine in 125ml glasses. On the small-serving nights, people drank around a third less over the course of the evening.

It could be that people count their drinks, or it's too much effort to keep going back to the bar, or that smaller glasses make drinking a smaller amount seem the right thing to do.

Whatever the reason, it's worth ordering the smallest measure or glass when you're out. And investing in smaller glasses at home too. I have been known to use sherry glasses for wine at home.

HOW TO DRINK THE WAY YOU WANT TO

Turn down the music

A French study showed that loud music in bars makes you drink more quickly. It may be that that you become more excited and drink more as a result, or that you drink instead of talking because you can't hear.[3]

Don't get involved in rounds or a kitty

Buy your own drinks; that way you will be in control. And you won't be tempted to drink to get your money's worth!

Use straight glasses

A study from the School of Experimental Psychology, University of Bristol showed that people drank lager 60 per cent more slowly out of straight glasses than out of outward-curving (pilsner type) ones. The theory is, you pace yourself while drinking. And you drink a curved glass faster because it makes you think the halfway point is lower than it really is.

In fact the same group showed that if you put markings on the curved glasses, people controlled their drinking better then too. The takeaway? Knowing how much you're drinking helps you regulate your drinking.

Talk to your GP

It's worth being honest with your doctor about how much you're drinking. They are not going to judge. The generally accepted rule of thumb among medics is that when you ask someone how much they drink, you then double their answer!

The doctor will use a brief questionnaire to work out how harmful your drinking is and whether you need specialist help or can cut down on your own (see page 149 for information on addiction). American research has shown that screening and one session of advice by your doctor can subsequently reduce the amount you drink at a single sitting by up to 25 per cent.

Surround yourself with non-drinkers

There are new kinds of gatherings specifically for people who don't want to drink. Try a morning rave, such as Morning Glory, or a no-drinking club, such as Club Soda. There are now even teetotal bars and pubs, such as Reformation. Just like a carnivore might eat vegetarian food some nights, you don't have to be 100 per cent teetotal.

Go out later

Research suggests that the later in the evening you start drinking, the less you will drink during the course of the night. You might expect that that late starters would try to catch up, but they don't. In a study from the Netherlands, people who started between 8 p.m. and 9 p.m. drank far more over the course of an evening than those who started later. The late starters were also less likely to binge.

If you're used to being at the bar on the dot of 7 p.m., plan to go out at eight thirty or even 9 p.m. Sure, some of your friends might already be plastered. But that might even put you off drinking too much![4]

No shots or cocktails

Buying a round of shots is never a good idea. They can contain from one unit up to three in one hit. Cocktails can contain even more.

One unit an hour

When you're planning how much to drink, you could think of it as one in, one out. On average, you process around one unit an hour. This strategy will take some sticking to, as it's only half a 175ml glass of wine or half a pint of beer or cider per hour. But there is another plus: drinking at this pace will likely help you avoid a hangover too.

Get an app

If you're the kind of person who likes to track your progress, there are plenty of apps that do this. Try: Drinkaware, Drink Free Days, DrinkCoach, Drink Less.

Don't drink at lunchtime

This is one of the easier rules to choose because it's getting increasingly unusual for lunchtime work functions to involve alcohol; I think because it's become clear that nobody does their best work when drunk, or even lightly inebriated. An alternative rule might be: never drink at work, including evening dos.

Weddings and similar parties can mean starting drinking early in the day, sometimes with very little food. But if the party is likely to go on until the evening, why don't you start drinking when the evening guests arrive? Or after you've eaten? Think about how many more interesting conversations you'll have – that you actually remember.

Don't drink on an empty stomach

Line your stomach, ideally with some protein, fat and wholegrain carbohydrate, and you'll slow down not only the rate at which you get drunk but also your peak blood-alcohol concentration levels – so you'll have less of a hangover too. Win-win!

Schedule an early-morning event

Want an excuse not to drink? Book in a chore, appointment or favour for the next morning for which you need to be awake, fully functioning and sober. Ideally, make it something where other people are relying on you so you're less likely to skip it, for example giving someone a lift or meeting for a gym class.

Don't use energy drinks as mixers

A 2016 study showed that caffeine makes you drink more, probably by increasing the rewarding properties of alcohol.[5] And a study the following year showed that energy drinks make you drink faster, too.[6]

Buy the third round

If you only want to drink two pints, buy the third round. That way you can order a non-alcoholic drink. You can even make sure it looks like alcohol so nobody can tell.

ESSENTIAL SAFER DRINKING RULES

Know your number

People need to know their alcohol intake in units the way they (should) know their calorie intake, their cholesterol level and their blood pressure.

Make your intake target three units a day or fewer

This level of intake is not 100 per cent 'safe' but represents a threshold of harms that most authorities consider an acceptable level in relation to the pleasures of alcohol. Above these levels, the harms of alcohol begin to accelerate as intake increases.

Take pride in lowering your number

Just as lower numbers for blood pressure and cholesterol give greater life expectancy, so lower alcohol intake leads to improved longevity and health.

Take two drink-free days per week

The past 50 years have seen a major change in the use of alcohol in the UK. Whereas in the 1950s alcohol was seen as a luxury, it is now widely perceived as a dietary staple and many people drink with every evening meal. There is evidence that short periods of not drinking, even just a day, can help the liver recover from the effects of alcohol and so reduce the risk of complications such as fatty liver and cirrhosis.[7]

WHY DRY JANUARY IS A GOOD IDEA

Taking a month off drinking has been shown to be a success for some people – and in real life, not just on social media.

It doesn't matter which month you choose – you could do Sober October or Dry July (although bear in mind that February is the shortest month!).

A survey by Dr Richard de Visser at the University of Sussex questioned people who did 2018's Dry January. It found that even when it got to August of that year, the participants were drinking for just 3.3 days a week, a drop from 4.3 days a week. Instead of getting drunk 3.4 times a month, they were down to 2.1 times a month. And they were drinking less on drinking days too: 7.1 units, down from their previous 8.6. Which is still way over the safe limits, but at least the trend is downwards.

Of course, this is only a survey, so there's a risk that it's based on people's edited and possibly inaccurate views of their month. That said, probably the most long-term and life-changing benefits were in people's reported feelings about alcohol. Nearly eight out of ten said they understood when they felt more tempted to drink, and seven out of ten had learned they did not need alcohol to have fun. Other reported benefits? Better sleep, more energy, a sense of achievement, weight loss and saving money.[8]

JUST SAY NO

If someone is used to you drinking, they might find it unsettling when you stop or cut down. So it can feel embarrassing and even insulting if you don't, for example, say yes to the wine at a dinner party, or to a round at the bar.

Because drinking is a social activity, peer pressure can be hard to resist. And that's not taking into account all the extreme drinking activities involved in joining some societies or how sports teams bond over drinking. Alcohol fuels that kind of group culture and vice versa.

I think drinking games – particularly hazing – should be banned. People die in drinking games. If someone wants you to take part in one, get away as fast as you can.

You shouldn't have to explain why you don't want a drink but you probably will. People don't realise the harms of alcohol so they still force drinks on others. But it's not really so different from forcing a cigarette on to someone, and nobody does that, do they?

Another reason people may insist is that they prefer those around them to drink so they feel their level of drinking is normal.

Prepare an explanation as to why you're not drinking, or not as much as usual. You could say: 'I don't want to get drunk because I don't want to embarrass myself.' Or: 'I am only drinking one glass because I believe that, for me, it's the right amount.' Or even: 'I feel good, I've reached the level of relaxation I was aiming for tonight.'

If you are an alcoholic, you could simply say that. This would be a very powerful statement indeed.

A softer way to approach not drinking is: 'I'm monitoring

my drinking.' After all, it's similar to monitoring calories and there's no shame in that. Or you could frame it in terms of calories, 'I'm trying to lose weight and I've used up my calorie allowance.'

If all else fails, and you're still being asked, follow the advice of this saying: 'No is a complete sentence.'

10

CHILDREN AND ALCOHOL

MOST TEENS START drinking around the age of 14 or 15, then more heavily at parties from the age of 15 onwards. The advice from the UK's Chief Medical Officer is that children shouldn't drink at all before the age of 15. And that aged 15 to 17, they shouldn't drink more than once a week. That's because evidence shows that alcohol affects the development of bones and hormones, as well as the brain and other organs.

There are two factors that I think really matter at this early point in a person's life. The first is you don't want your child to be a statistic. Young people do die of alcohol poisoning. That means teaching them how to be safe if they are going to drink now. The second is about the long term. If your child does decide to drink, you want them to learn a sensible way of drinking that will mean they don't harm their health in the long term.

It's true that some teens don't use alcohol at all now, which must be a relief to parents. But that's not true of all teens: WHO's *Global Status Report on Alcohol and Health 2018* showed that in Europe, 44 per cent of 15- to 19-year-olds do drink.

If your teen wants to drink, you are unlikely to be able to stop them having any alcohol. More worryingly, nearly one in five 15-year-olds think it's OK to get drunk once a week.

People often ask me if they should let their child drink at home, or at a party. That is a hard question. If you have or have had an alcohol problem yourself, I'd say the answer is probably no. They may have a similar genetic or other predisposition to alcoholism as you.

Otherwise, introducing alcohol – in the form of beer or cider of a strength less than 4 per cent, or very diluted wine – on special occasions at home from the age of 15 is probably fine, although it's a myth that it will stop your child binge-drinking. Bear in mind that statistics show the earlier your child drinks, the more likely they will have problems with alcohol later in life (see figure on page 219). Of course this doesn't prove that it is drinking alcohol earlier that leads to dependence, as there may be genetic or environmental factors, but it is worth knowing.

Even if you don't have an alcohol problem, think about what your own drinking is saying to your child: is your habit of finishing a bottle of wine a night giving the opposite message to your lecture on moderation? Has your child seen you drinking to get drunk? Of course children will do what you do, not what you say. And don't be the person who buys them drink.

I think the secret to helping your child drink sensibly – as well as drinking sensibly yourself – is education. Tell them how to use alcohol safely.

Explain that people can and do die from alcohol poisoning after just one night's drinking. And the reason why we don't always hear about it in the news is that it is common enough not to be news. After Leah Betts died in 1995 after taking ecstasy at her eighteenth birthday party (in fact she died of water intoxication), her picture was on billboards all over the country. But when, in 2008, Gavin Britton, aged 18, died during an 'initiation' for the golf society at Exeter University, from drinking shots and spirits, his wasn't.

Fig 6. Prevalence of lifetime alcohol dependence according to age of drinking onset

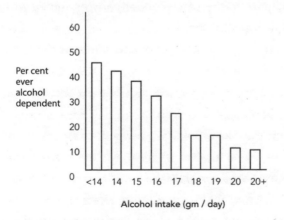

Per cent ever alcohol dependent

Alcohol intake (gm / day)

You can explain that getting drunk is more likely to lead to fights, accidents, injuries, unwanted sexual experiences and being mugged, as well as to doing things they will regret such as the Sheffield Hallam University student who was filmed peeing on the wreaths of a war memorial during a Carnage UK freshers' event and was lucky not to be sent to prison.[1]

HOW TO HELP TEENS DRINK SAFELY AT A PARTY

If your teen is having a party at home, the key is supervise, supervise, supervise. If your child is going to a party at another child's house, can you speak to the parents and make sure they are going to be there?

Bear in mind that this advice is for age 15 plus. Before that, I'd advise no alcohol at all at parties.

- Don't go out. Be obviously present. Do not trust teenage children to the care of older children – they won't take any notice of them.
- Confiscate any strong alcohol and spirits as the teens come in.
- Confiscate any drugs. My message re drink and drugs is very simple: Don't drink and drug.
- If any child looks drunk, take away their alcohol, put them somewhere quiet where they can sit down and give them some water. Make sure someone is looking after them and that they get home safely.
- If any child passes out and can't be roused, put them in the recovery position and call an ambulance.

HOW TO TALK TO YOUR CHILDREN ABOUT BOOZE

Alcohol should be a sensible conversation throughout your child's life, not just when they ask to have their first party with alcohol. Obviously what you say needs to be made age-appropriate: the following list is aimed at a 15-year-old.

1) **IT WON'T MAKE YOU HAPPY** Alcohol affects your mental health – it can lead to anxiety and depression, or make them worse. A survey by NHS Digital, *Smoking, Drinking and Drug Use among Young People 2018*, showed that a third of young people aged 11 to 15 who'd recently drunk alcohol had low levels of happiness. For those who hadn't recently drunk, it was 22 per cent.[2]

2) **AVOID SPIRITS** You won't have a sense of how to regulate your drinking yet. So it's way better to drink weak beer than strong spirits, in particular shots. With spirits, you can get catastrophically drunk before you've even realised it. In fact, it's a good idea to avoid any alcohol that is more than 4% ABV.

3) **YOU CAN DIE FROM ALCOHOL POISONING** According to the Alcohol Education Trust, for 16–24-year olds, 21 per cent of deaths in males and 9 per cent of deaths in females can be attributed to alcohol consumption.[3] For example, first-year student Ed Farmer died from drinking at an agricultural society initiation in 2016 at Newcastle University, after drinking shots.[4]

4) **BE AWARE YOU'RE BEING SOLD ALCOHOL** Drinks advertising is clever at seducing you into wanting to drink, by associating drinking with sex, money and fun. The drinks industry wants you to start drinking. And

research suggests that the more alcohol advertising teenagers see, the more they drink. Your favourite sport is likely to have an alcohol brand as a sponsor – Budweiser sponsors the Premier League, for example.[5] And the same might be true of the last event or concert you went to. The hospitality industry – aka the alcohol industry – also sponsors sporting and cultural events and invites influential people, from celebrities to politicians to journalists.

5) HAVE ONE DRINK (OR TWO MAX) If you have one drink, you'll have fun but you shouldn't get drunk (if it's 4 per cent beer or cider). If you have two, you may be drunk.

6) IF YOUR FRIEND IS DRUNK, HELP THEM Don't do what the friend of Tony Blair's son Euan is said to have done – assume he'd gone home when in fact, he was lying unconscious in London's Leicester Square.[6] Make sure you get your friend home. Don't be frightened about telling their parents – they will very likely be grateful to you. They'd rather their child was safe.

Your friend may not just be asleep. If they are unconscious and unrousable, call an ambulance. Don't do what many so-called friends do and make fun of them or do horrible things to them.

7) DRUNK PEOPLE GET INTO TROUBLE They have accidents and fights, commit criminal acts, get arrested. One doctor I know, when she was a medical student, woke up in a stranger's house. The man explained she had knocked on his door the night before, so drunk she couldn't speak. Having no idea who she was, he had let her sleep on his sofa. She'd gone home to the wrong house. She couldn't remember any of this. She was lucky she'd stayed safe.

8) GET HOME SAFELY You are vulnerable after drinking because you are less aware of, for example, a mugger coming up behind you and less likely to recognise or care about danger. Stay with your group of friends. Don't go home alone. Don't get into a stranger's car. Don't get an unlicensed minicab. Don't walk home if it's not safe or you wouldn't do it sober. Work out how you'll get home before you go out.

9) DON'T DRINK AND DRUG Never, ever take drugs when you are drinking. One, you can't predict the result of mixing the two. And two, you may not know what you're taking or you won't be aware enough to be certain.

In 2010, two boys in Scunthorpe, Louis Wainwright, 18, and Nicholas Smith, 19, were reported to have died of mephedrone aka meow meow, an Israeli drug that was originally created for controlling aphids.

CNN called me for a comment on the case; I said I was very surprised at this because in an Israeli review of 400,000 people who'd taken this stimulant there had been no deaths.[7]

Later, the real story came out. The boys had been drinking all evening in several bars, then when the bars shut had gone into the town to buy drugs. It turned out that instead of mephedrone, they had bought and taken the heroin substitute methadone. It might have been that they were so drunk they didn't know what they took. Also, if you take drugs when drunk, you're more likely to die. Methadone and all other opioid drugs suppress breathing and alcohol accentuates that effect.

DRINK?

10) DON'T DRIVE WHEN DRINKING This should go without saying. Hundreds die every year from drink-driving accidents in the UK, thousands globally. Other things not to do if you've had a drink: climbing and/or jumping off high things, especially into the sea from a pier or into a river or reservoir or pool. In fact, don't go swimming. And never give or take dares.

11) WEAR CLOTHES As alcohol dilates (widens) your blood vessels, it makes you feel warm even when it's cold outside. So even in winter, people who've been drinking will go out without a coat. But when you are drinking you also lose heat faster, so it's easier to freeze to death – especially if you sit down and then fall asleep outside. So wear proper clothes. And don't ever stop and sit down on the way home – especially not in a snowdrift.

12) LEARN FROM YOUR MISTAKES If your friends tell you what an idiot you were last night, believe them and learn from it. You may not remember – be thankful that someone does and is watching out for you.

13) DON'T POST OR MESSAGE Or email or text or go on social media. You will likely regret it.

14) WALK AWAY FROM FIGHTS People can get incredibly violent when drunk and really hurt each other. They get disinhibited and easy to offend. I have seen a fight where people were glassed. I hid under a table, watching the blood pour out of them. If a fight starts, walk – or run – away.

15) KNOW WHAT MAKES FOR CONSENT It is not possible to give consent while you are drunk. So don't have sex with anyone who is drunk (or who is under the age of 16).

CHILDREN AND ALCOHOL

One in three women have been sexually taken advantage of while under the influence of drugs or alcohol, mostly in private homes by someone they know, according to the 2019 *Global Drug Survey*. Nearly 90 per cent of these cases involved alcohol.

11

IS THERE A SOLUTION TO THE BOOZE CRISIS?

WHAT SHOULD SOCIETY DO ABOUT THE BOOZE PROBLEM? AND HOW DOES IT AFFECT YOU?

SCIENTISTS HAVE A word to describe the fact that we live surrounded by available and attractive food that conspires to make us eat; they call it our 'obesogenic' environment. I think the same is true of alcohol: the environment we live in is alcogenic. It's far too easy to drink too much.

At the moment, if you want to drink less, it's as if you've decided to stop gambling but you live in a Las Vegas hotel. There are massive shelves of booze in our supermarkets, adverts for drinking that make it look cool and fun; there are

226

bars and pubs on every high street, and most of our British social life takes place with the help of booze.

Despite repeating their message of 'sensible drinking' at every opportunity, the drinks industry wants you to keep drinking. That is its reason for existing. One estimate is that, if everyone drank within recommended limits, the industry would lose £13 billion. That's a lot of lost profit. Its aim is often aided by the government, which wants the tax income.[1]

Over the years I worked as an expert advisor to the government in this area, I saw politicians repeatedly talk up how they were going to solve the 'alcohol problem'. But when it came to actually adopting any useful policies that might make a difference to the crime and health harms caused by alcohol, they rarely did. And when it came to dealing with an advisor – me – who spoke up to tell them this, their solution was to sack me.

In this chapter, I'm going to recount my experiences and frustrations of working as an advisor. And I'm going to suggest the measures that have worked around the world, and which I think could help create an environment in the UK where fewer people – including you – end up drinking too much.

To analyse what's happening with the drinks industry now, it's enlightening to look back at what the tobacco industry did in the 1950s and 60s. It spent decades lying about the link between smoking and lung cancer, and denying evidence of any health harms. It also tried to deny the fact that tobacco was addictive.[2]

In a way, you could say the government of the time colluded with the industry. In 1956, the first British epidemiological study, by Richard Doll and Austin Bradford Hill,[3] showed a causal link between tobacco and lung cancer and heart attacks. Learning about this in the early 1970s, as an undergraduate studying medicine at Cambridge, turned out to be my first exposure to the interaction of drugs and politics. One of the senior science tutors told me that in the 1950s he'd been on the government committee set up to work out what to do with the Doll discovery. He told me how illuminating it was to find out that smoking was taking 15 to 20 years off the life of men in the UK. He also told me the committee realised that if they stopped people smoking, it would result in an enormous extra burden of elderly people on the health service and pension system.

After all, in terms of cost to the country, smokers are good value. They pay a huge amount of tax via duty and during their working lives, but many die before they can draw their pension. The cheapest death for the NHS is a heart attack. So, for the good of the national finances, the committee decided not to act on the study.

(I'd like to add a caveat here: dying from cigarettes is by no means always quick and easy. As a medical student working in a men's chest ward in 1973, 29 out of 30 beds were smokers dying of lung cancer. And lung cancer is a horrible way to die.)

Eventually, of course, the truth about tobacco was made public, and this made the tobacco industry look terrible. I think the drinks industry learned several lessons from Big Tobacco. Firstly, they learned not to lie. They didn't say that alcohol isn't addictive or that it doesn't cause health harms. That would have been absurd.

Instead, they took the line of 'drink responsibly'. What this means in reality is unclear. I presume it means drink within the recommended limits, so drinkers will be exposed to a lesser degree of harm. In other words, they put the blame for any problems caused by alcohol on the 'irresponsible' drinker and not on the drink. But as they well know, one of the cardinal effects of alcohol is to dissolve responsibility. And that's not even taking into consideration the up to 15 per cent of drinkers who become dependent.

THE 1990S AND BIG ALCOHOL

When Tony Blair came into government in 1997, he announced that he was going to have a critical look at drugs and drug addiction. He said: 'I want to breathe new life into the battle against drugs. We will hit hard on drugs and the drugs trade.'[4]

He set up a cabinet office committee to tackle this. I was asked to join. And so I took the early train up to London every other Monday morning for the meetings.

We produced a super report. But when it came to its publication I discovered – with horror – that the chapter on alcohol had been removed. The scientists on the committee protested and were told by the secretariat that they had consulted the drinks industry and had been told that some of our science was wrong.

They had also promoted the idea that alcohol has health benefits and so shouldn't be considered alongside other drugs. Instead of referring back to us, the government decided to drop the chapter! (NOTE – I am not sure they ever released the original report.)

We protested. The government then decided to publish a separate report on alcohol alone. Their report – which didn't come out until 2004 – was called *UK Cabinet Office (2004) Alcohol Harm Reduction Strategy for England*. Read this along with the thoughtful critique by Professor Robin Room.[5]

In the meantime, many of the health scientists from the original committee got together and worked on an independent report under the Academy of Medical Sciences – the leading independent UK academic medical authority, of which I am a Fellow. Our report, *Calling Time: The Nation's Drinking as a Major Health Issue*, also came out in 2004.[6]

Surprise, surprise – the two reports were at odds with each other. The government report looked as if it had been written by the drinks industry lobbyists. It was lacking the effective policies we recommended: banning TV advertising and sports sponsorship and making alcohol duty reflect actual alcohol content to bring up the price of super-strong cheap cider, for example; and adding health warnings to labels. And when it came to parliament, guess which report won out? If you want to read more about the whole sorry saga of political machinations overruling evidence-based public health measures, it's eloquently described in this piece by the world's leading alcohol policy expert and a great teacher of mine, Professor Robin Room.[7]

In 2000, I'd joined the Advisory Council on the Misuse of Drugs (ACMD) as a scientific advisor. It's a public body set up to monitor the application of the Misuse of Drugs Act

1971, to advise the government on drugs policy as well as classification. My role was to chair the subcommittee that explored in detail the pharmacology and toxicity of new and old drugs and advise the full Council on health threats and policy options, especially for new entrants to the drug scene.

My whole career, I've lobbied for a rational policy on drugs, including alcohol. When I took on the role of Chair, I made it a condition that I'd be allowed to develop more scientific and transparent methods for assessing the harms of drugs. The point I always made and still make is that drugs should be classified according to the harm they do, not by how they are perceived by the voting public. And assessment of harm should be measured scientifically, not according to a knee-jerk reaction.

For the next nine years, we did exactly this, producing landmark papers in *The Lancet* in 2007 and 2010. Despite being told by the Home Office that we weren't allowed to consider alcohol (or tobacco) as drugs, we persisted in including them. And in the process, we discovered that alcohol was much more harmful than most people, including me, had supposed. It became obvious that the government's policy on drug classification was often inconsistent with the harms each drug actually does. The fact that I continued to consider alcohol to be a drug turned out to be a major factor in my sacking later on.

In the first systematic analysis of drug harms, published in *The Lancet* in 2007, we included what you'd expect – illegal drugs such as heroin, cocaine, cannabis – but also some legal drugs such as alcohol, tobacco and solvents. And we assessed all the possible harms of each substance, including likelihood of dependence, the physical harm to the person using it and the effects on the people around the user and on society.

The report placed alcohol as the fifth most harmful drug to the user after heroin, cocaine, barbiturates and methadone. Tobacco ranked ninth. Cannabis, LSD and ecstasy, while still harmful, ranked lower, at 11, 14 and 18.

I was so good at my role chairing the scientific committee that, in 2008, I was promoted to Chair of the full Council, and so became government chief drugs advisor, nicknamed the 'drugs czar'.

POLICY VS SCIENCE

My first big run-in with authority was early in 2009, with Jacqui Smith, then Home Secretary. In a speech, I compared the 100 deaths a year caused by horse-riding to the 30 deaths a year linked to ecstasy. This may sound like a weird comparison, but the point I was making was that we accept the status quo without question, all the dangerous things we've always done, including both horse riding and drinking. But any new drug – at that time it was ecstasy – is banned as it's thought to be too dangerous. When you look at the evidence, this is illogical.

It was a talk I gave on the results of the 2007 study nine months later that led to the then Home Secretary Alan Johnson asking me to leave the ACMD. Specifically, he objected to my comments that cannabis was being demonised out of all proportion to its risks when alcohol was considerably more harmful. He said I couldn't be both a government advisor and a campaigner against government policy. My reply was, that's what expert scientific advisors were bound to do if government policy goes against scientific evidence!

After I left the ACMD, I completed my second report ranking drugs, published in 2010. It assessed 20 drugs using 16 criteria, nine related to the harms that a drug produces in the individual and seven to the harms to others.

It showed that heroin, crack cocaine and methamphetamine were the most harmful drugs to individuals, whereas alcohol, heroin and crack cocaine were the most harmful to others. Overall, alcohol was the most harmful drug (overall harm score 72), with heroin (55) and crack cocaine (54) in second and third places.

When I talk about these results, people often ask if it's fair to compare illegal with legal drugs. They correctly point out that alcohol causes more harm because it's legal and therefore much more widely used.

My answer is, we took this into account in the analysis. We weren't saying that alcohol is the most harmful drug to users (that accolade went to heroin). But because of alcohol's wide use, it scored as the most harmful to others (and, I'd say, would still do so now). So, when we combined harms to users and society, it came top. And remember, the cost of all alcohol harms are largely picked up by others – that is, you and me. The logical conclusion is, if government drugs policy is about harms, alcohol should be the primary focus. But for political reasons, this evidence has been ignored.

HOW POLICIES AND HARMS INTERACT

THE IMPACT OF 40 YEARS OF ALCOHOL POLICIES ON DEATH RATES

The graph opposite depicts what I think is probably the most severe failure ever of UK health policy. The lines show changes in mortality rates from a range of different causes over 40 years.

The data is corrected and standardised so at the start they are all at 100 per cent. That allows changes over time to be easily seen as an upward (increase) or downward (decrease) slope of the lines.

What you can see is that death rates from (almost) every disorder have gone down since 1970. This is what you'd expect, as medicine improves and society becomes generally healthier due to public health measures. In fact, most disease rates have halved over the 40-year period.

But as you can also see, deaths from liver diseases (80 per cent of which are caused by alcohol) have increased fivefold.

Successive governments have had access to this data. I can only assume they all decided not to take any steps to cut the rate of liver disease because they didn't want to stand up to the drinks industry.

Instead, what they did or didn't do compounded the problem. Firstly, they didn't ensure alcohol prices kept pace with income because they refused to raise duty enough or introduce minimum pricing. In real terms, alcohol now costs a third of what it did in 1970. Secondly, they liberalised access to alcohol by allowing sales in supermarkets and by lengthening licensing hours.

The upshot was, over those 40 years, the average per-head consumption of alcohol increased by about 50 per cent. But because of the exponential relation between consumption and liver damage, the number of deaths multiplied by 400 per cent.

For over 40 years alcohol and health experts have been clamouring for governments to take notice and reverse these damaging policies. But our voices have not been heard; governments of both colours like alcohol too much.

Fig 7. A comparison of death rates from liver and other diseases over the past 50 years

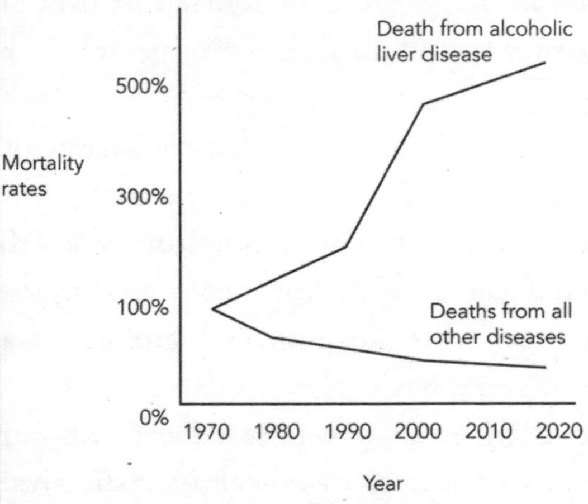

Source: Figure courtesy of Professor Nick Sheron, personal communication.

MAKE THE PRICE RIGHT

Alcohol needs to be made more expensive. It's been shown that when the price goes up, consumption goes down.[8]

We should start by taxing drinks on the amount of alcohol they actually contain. At the moment, the duty system doesn't make sense. Why should the tax on alcohol in a cider containing 10 per cent of alcohol be one-third that of a wine of the same strength? All that does is encourage the consumption of strong ciders. We should also cut the availability of super-cheap booze, including those high-strength ciders. And we should stop supermarkets using discounted offers on alcohol as loss leaders. Does it really make sense for alcohol to be sold at less than the price of water?

The price measure I favour, though, is the introduction of minimum unit pricing (MUP), as happened in Scotland in 2018. The law there says that the minimum price per unit must be 50p (you can charge more, of course). At the time the law was passed, over half of all alcohol was being sold below this price level.[9]

There have been plenty of arguments against this policy; the main one being that it is unfair on most drinkers. But neither you nor I are drinking alcohol that's less than 50p a unit. When did you last buy a bottle of wine that was less than £4.50? The market for cheap alcohol is mostly made up of teenagers and alcoholics.

Why does MUP work so well? It reduces consumption; but the real beauty of it is its non-linearity. Estimates from the Sheffield Alcohol Research Group that has modelled the value of minimum pricing suggest that a 50p unit price would lead to at least a 10 per cent reduction in consumption.

IS THERE A SOLUTION TO THE BOOZE CRISIS?

As I explained in Chapter Two, the relationship between alcohol intake and harm shows an exponential curve; drinking a litre of wine each day is five times more harmful than drinking half a litre, not twice. As the heaviest drinkers contribute the vast majority of the health costs, decreasing their consumption has the biggest impact.[10] If people who drink a litre of wine a day (or its equivalents) drank 10 per cent less, it would reduce harms by at least 25 per cent – see the graph in Chapter Two. The Institute of Fiscal Studies has recently confirmed the value of this targeted approach.[11]

Another benefit of minimum unit pricing is that it largely affects sales in supermarkets rather than bars and pubs, so protects the latter from more closures.

When David Cameron came into power in 2010, he said he wanted to reduce the problems of alcohol. And one of the measures he supported was MUP. So the Department of Health set up the Responsibility Deal Alcohol Network (RDAN), which was designed to bring together the drinks industry, local authorities and health experts to encourage drinking within recommended guidelines. The problem was, the government decided this group was to be made up of 50 per cent health experts and 50 per cent representatives of the drinks industry.

At first there was an attempt to work together. But in 2011 all the health experts resigned after the industry representatives blocked all suggestions of compulsory policies such as minimum unit pricing. It was clearly impossible to be objective when half the committee had a specific link to the industry that was causing the problem![12]

This committee is now defunct. In fact, there is currently not a single body in Britain that is empowered to argue the case for sensible alcohol policies.

The government doesn't want to mess with the alcohol market in England because they make so much money from it. But in fact, it's been estimated that when you add in the costs of alcohol to society, there is a net loss to the Exchequer.

This is undeniably a difficult argument to disentangle economically, and a complicated sum. But the costs of alcohol to society are relatively well established. These are: £3.5bn on health, especially hospital admissions and accident and emergency attendances; £6.5bn for policing drunkenness; £20bn for lost productivity through hangovers. The total is £30 billion.

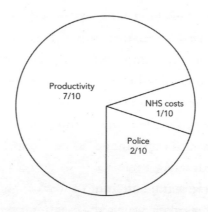

Fig 8. This figure shows the relationship between personal income and alcohol consumption across the world

Source: http://www.alicerap.eu
With permission from Prof Jurgen Rhem.

Another estimate of the costs to society comes from the National Social Marketing Centre. This has a wider remit including lost income due to unemployment, the costs to social care, criminal justice and fire services and the pain and grief associated with illness, disability and death. Their total figure is £55.1 billion.

Taxation of alcohol raises about £20 billion a year. That leaves a net deficit of up to £30 billion.

But the government are terrified of making any change to alcohol's status because its income from taxation is immediate. The industry pays its tax quarterly and duty is coming in constantly. The Treasury view is that if things changed, the tax income would reduce but the health benefits wouldn't happen for 10 to 20 years.

The truth is, some of the health costs are immediate. Think about an A&E department on a Friday night: if it was only half full, you wouldn't need so many nurses and doctors.

And if you can reduce someone's consumption of cheap cider from, say, two litres a day down to one, they can then be treated more cheaply in a normal ward and not in intensive care, as they won't be dying. These are immediate and significant reductions in health costs that the Treasury have overlooked. And there would be savings in the £6.5bn yearly policing costs for drunkenness too.

Going back to the techniques that Big Alcohol has used to keep on growing as a business, I'd say it has learned to dissemble. Some years ago, I remember listening to a discussion on the *Today* programme on Radio 4.

The first speaker was an academic who had used the Sheffield Alcohol Policy Model (SAPM) to assess the effects of minimum pricing on different socio-economic groups, including health harms. He said the model showed that MUP would cut consumption.

There was a contributor from the drinks industry on too. He simply said: there is no proof minimum pricing will work in the UK.

I could hear the cogs working in the head of the academic. He started to stutter. Because there is no arguing with this kind of non-logic. There is never any 'proof' until you do the experiment or introduce the policy – that is the reason we use models.

I rehearsed what I would have replied. I'd have said: all the modelling that's been done shows minimum pricing is likely to work. A review in the *BMJ* looked at over 500 studies and concluded: 'price-based alcohol policy interventions such as MUP are likely to reduce alcohol consumption, alcohol-related morbidity and mortality.'[13]

And if you don't try, you won't find out.

Using price to control alcohol consumption is not a new policy: in fact, we have done this for over 200 years. It is the original principle behind duty, which was based on the strength of alcohol. That's why the duty on spirits is higher than that on beer.

Except there's a flaw in duty, now: super-strong ciders and beers. Although the duty does go up in bands according to strength, both beer and cider still pay way less duty per unit than spirits and wine. This makes extra-strength cider the cheapest way to get drunk.

For example, a 2.5 litre bottle of 7.5 per cent cider – say Frosty Jack – contains nearly 20 (19.8) units of alcohol, almost a week and a half's units in one bottle. It costs £3.70 in a supermarket. The duty on this is 50p a litre, so £1.25 a bottle. You are paying 6p duty per unit.

Now look at a 750ml bottle of 13.5 per cent wine, for example Campo Viejo Rioja Tempranillo, which is on sale in a supermarket for £8. The duty on this is 298 pence per litre, so £2.23 duty on 11 units. You are paying 20p duty per unit.[14,15]

This figure from a new report from the Institute for Fiscal Studies shows the anomalies in current taxation of different forms of alcohol.[16]

The graph shows the current UK tax situation on different sorts of alcohol in relation to the amount of alcohol in each one. For spirits, the alcohol content is always the same and the tax is constant too. But for the other drinks, tax is levied on the type of drink rather than the alcohol content. You can see that each type of alcohol has a different range of taxations and these vary with the strength of alcohol. So for cider the amount of tax on the alcohol gets proportionately lower the stronger the cider. Whereas for wine, the tax per unit of alcohol fluctuates wildly, after 8 per cent dropping rapidly downwards as the alcohol content goes up.

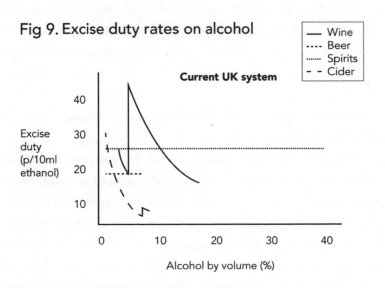

Fig 9. Excise duty rates on alcohol

An optimal tax system is where all types of drink are taxed for the amount of alcohol they contain. It includes a rise in tax for spirits over 20 per cent alcohol, as these are the biggest contributors to the societal costs of alcohol.

THE SCOTTISH PRICE WARS

In 2012, the Scottish Parliament voted to bring in minimum pricing. It's been fascinating to watch what has happened since. It shows not only the power of the drinks industry, but also that minimum pricing can work.

The Scottish government made this decision because they realised that alcohol was destroying communities. Scotland had (and has) by far the highest level of health harms from alcohol in the UK. For example, over one-quarter of intensive-care beds are filled with people dying of alcohol-related illnesses such as liver and heart disease.[17]

What happened next was salutary. The beer and spirits industry – with the exception of Tennents, a Scottish company – funded a legal challenge against minimum pricing, which was pursued by the Scotch Whisky Association.

It turned into an extremely long-running legal challenge. First, the Scottish Court of Session found in favour of the government. The alcohol consortium kept going. It said that minimum pricing wouldn't work, that it would penalise the poor and that there were other ways to reduce the harms of alcohol. They also argued that it would go against European Union trade rules. The European Court of Justice disagreed and sent it back to the UK.

Finally in November 2017, the UK Supreme Court ruled

that MUP was legal. And so finally, on 1 May 2018, minimum pricing came into force, at 50p a unit (two units is one pint of beer or one 175ml glass of wine).

First data from Scotland looks pretty promising. Research from Newcastle University has shown that the amount of alcohol bought in shops and supermarkets per person per week fell by 1.2 units (just over half a pint of beer or a measure of spirits) compared with what would have been drunk without MUP. In England over the same time, consumption increased.[18]

The biggest drop – two units a week – was in the heaviest fifth of drinkers. This shows that the policy is having the biggest impact on exactly the people it needs to. The number of units bought per person per week in Scotland is still higher than in England, at 19 units, but it did start from a much higher level. The Welsh Assembly has now passed legislation for MUP.

At the same time as MUP was introduced, Scotland made two other important policy changes for alcohol. It reduced the drink-driving limit (from 80mg% to 50mg%) and banned discounted alcohol such as bulk-buys in supermarkets and two-for-one offers in bars. All the evidence suggests that, together with MUP, these should result in a real cut in harms.[19]

Proper policies on alcohol can lead to a virtuous circle of health and wealth. France is a remarkable example of this. It shows you can have an alcohol market that's just as economically successful as a non-price-restricted one but which has a massively reduced level of harm.

THE FRENCH EXAMPLE

In my lifetime, the French government have got policy on alcohol right while, in the UK, we've mostly got it wrong. This is a useful comparison, as France isn't a million miles away and as a nation they're not that different, genetically.

People in France do still drink over the recommended units. But they drink a lot less than they used to. When I was a student, it was rare to see someone in hospital in the UK with liver cirrhosis. But not in France, which is why we called it the French disease. Now, French cirrhosis rates are lower than the UK's, because their government has acted in the best interests of their people while ours has been suckered by the drinks industry.

In the 1980s the French government did a really detailed analysis of drinking in France. They reached the conclusion – and it's the only one they could have reached – that it was a serious health problem. But, as in the UK, there were competing interests. Forty per cent of French citizens got some kind of monetary value from the drinks industry. But even taking this into consideration, the experts realised that excessive alcohol intake was uneconomic.

The government tackled four key areas: advertising, health warnings, price and drink-driving.

1. ADVERTISING The 1991 *Loi Évin* – the Évin Law – takes its name from Claude Évin, who was then minister for health. The law banned most alcohol advertising on TV and in cinemas, and all alcohol ads aimed at young people. They decided advertising content must be factual, not aspirational. They stopped alcohol sponsorship for cinemas, festivals and sporting events. (By the way, seven years later France won the FIFA World Cup for the first time and in 1999 they got into the finals of the Rugby World Cup, which shows teams don't need the support of the alcohol industry to be good at sport.)

2. HEALTH WARNINGS The law put the health warning 'alcohol abuse is dangerous for health' on every label. And also: 'the consumption of alcoholic drinks during pregnancy, even in small amounts, can seriously damage the child's health'.

3. PRICE The government placed restrictions on happy hours and stopped alcohol being given away. Also, most alcohol sold in France is wine, and the government priced out cheap alcohol by pressuring the industry to make quality wine.[20]

4. DRIVING In 1996, the government also reduced the BAC for driving, down from 80g% to 50g%.[21]

The results? There has been a dramatic reduction in road deaths and people sustaining serious brain damage from accidents. Awareness of the dangers of alcohol to the foetus has gone up and drinking in pregnancy gone down.[22] I predict that this will lead to a reduced rate of foetal alcohol syndrome in the coming generations.

In the 1970s, France had death rates from liver cirrhosis over five times greater than the UK; these have now dropped by half. In the UK, ours are now as high (and in Scotland higher) than in France.

Fig 10. Liver mortality in the UK and France

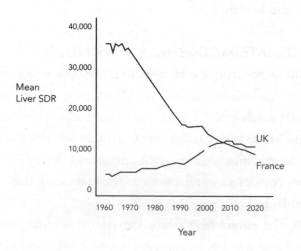

Source: https://www.journal-of-hepatology.eu/article/S0168-8278(18)32057-9/pdf

The figure shows the remarkable fall in standardised death rates (SDRs) from alcoholic liver disease over the past 50 years in France as they cut their consumption by half over this period. In stark contrast over the same period UK death rates nearly tripled as our alcohol consumption nearly doubled.

The French policies turned out to be a win-win for the wine industry too because there is more profit in making quality wine. And that's despite the fact that people drink less of it. As someone once said to me, 'I've never seen anyone get drunk on a 1961 Château Latour.'

The UK drinks industry know this data. They know that if we went down the route of raising the price of alcohol, they would be more profitable over time. In my opinion, the reason they don't is that making money in the short term is so easy.

And they think that if they make any concessions in this direction, people will begin to question all our other beliefs around alcohol and health.

THE ULTIMATE ALCOGENIC ENVIRONMENT?

You could say alcohol oils the wheels of government. In the Palace of Westminster, drink is subsidised and there are nearly 30 places to drink.[23]

In 2012 Eric Joyce, Labour MP for Falkirk West, headbutted a Conservative MP, Stuart Andrew, in the Strangers' Bar. He was convicted of four common assaults and so was expelled from his party.

In 2007, Labour MP Fiona Jones died of alcoholic liver disease.[24]

Charles Kennedy, the former Liberal Democrat leader – who bravely spoke out about his alcoholism – eventually died from a major haemorrhage that was a consequence of liver damage.

Drinking on an official trip led to the death of the Labour MP Jim Dobbin, who, according to his wife, only drank socially. He died in 2014 in Poland, after a dinner at which he drank a shot of Polish spirits with each course. His blood alcohol level at post-mortem was 399mg per 100ml, nearly five times the UK drink drive limit.[25]

Drinking in parliament is rife.[26] Many MPs have admitted to being affected and one has even described being too incapacitated by alcohol to be able to walk through the voting chambers.[27] I have argued that to allow intoxicated MPs to make decisions on issues of national importance such as going to war is outrageous. In fact, I believe MPs should be breathalysed before being allowed to vote.

The government has a long and close relationship with the alcohol industry – a special relationship, I'd call it. The club with the most members is the All-Party Parliamentary Beer Group. It's generally believed that over half of all MPs belong to it.[28]

In recent years, the group has helped the beer industry lobby for both cuts and freezes in beer duty. It also gives out a couple of cases of free beer to each member every Christmas.[29] MPs have to declare when they've been given freebies; the list always includes invitations to alcohol-industry-sponsored 'hospitality events'.

That's not to say all MPs are in favour of alcohol advertising and sponsorship. During the coalition government, we (the health experts) recommended curbs on alcohol advertising and sponsorship for the UK, as well as MUP. In 2012 the Health Select Committee, headed up by Stephen Dorrell, a former health secretary, called for a British equivalent of the *Loi Évin*. It recommended a series of ranked proposals for how to deal with the problems of alcohol, naming changes in price, advertising and promotions as the most effective strategies.[30]

What happened was that the alcohol industry persuaded the government to reverse the order and to only bring in the least effective of all these measures. That measure was the introduction of warnings on bottles that said: drink responsibly.

The industry also agreed to avoid having images of people under 25 years of age drinking in their adverts. But that didn't and won't stop alcohol infiltrating youth culture. A US study found that an average of 22.4 per cent of songs in the Billboard Hot 100 from 2007 to 2016 mentioned alcohol. And since the rise of social media, it's been easy for the industry to achieve high levels of product recognition in young people.[31] As a report from ALICE RAP (Addiction and Lifestyles in Contemporary Europe – Reframing Addictions Project), a five-year European research project of 200 scientists from 29 different disciplines says: 'Even simply watching a one-hour movie with a greater number of drinking scenes, or viewing simple TV advertisements, can double the amount drunk over the hour's viewing period.'[32]

The outcome of the select committee was a blatant and outrageous reversal of what should have happened. When it comes to deciding how to prevent the harms of alcohol, the drinks industry doesn't even have a right to be in the room. You might well ask why they were there. The truth is that they (and their lobbyists) persuaded government that they were THE experts on alcohol and would work with health experts to find solutions. But their solutions are always the most industry-friendly – which is why nothing has changed in the past decade.

WHAT WORKS WELL IN SWEDEN . . .

I am in favour of the Swedish model of alcohol control, a form of state control called Systembolaget. This is a chain of government-owned shops, the only places allowed to sell alcohol of more than 3.5% ABV.[33]

When you go into a Swedish state alcohol supermarket, it looks like the biggest duty-free shop you've ever seen. There are thousands of different versions of alcohol for sale, including an enormous amount of high-quality wine. This makes the government effectively the largest wine bar in the world.

There is only one shop per town, and it's only open 9 a.m. to 6 p.m., Monday to Friday, as well as Saturday morning. There are no 'buy one get one free' promotions allowed. And you have to be 20 to shop there (although you can drink in restaurants and bars from the age of 18).

The upshot is, you need to plan if you want to buy alcohol. It makes that purchase much more conscious.

This system has led to the Swedes drinking roughly three-quarters of what we drink per year in the UK. And to a rate of deaths from liver cirrhosis approximately half of ours. This shows, yet again, how small changes in consumption lead to bigger health gains.[34,35]

In 2010, I set up the charity DrugScience, to tell the truth about alcohol and other drugs. In 2018, we got a grant from the Norwegian government to work out the best possible scenario for how drugs might be sold and regulated.

Our research team used a new version of the same process used in my two previous papers to estimate drugs harms, but this time looked at policies and benefits. We looked at four very different scenarios, for both alcohol and cannabis, and at the impact they would have on the user and society, from health to education to social behaviour to policing and economics. The four were: 1) state control as in the Swedish alcohol model, 2) prohibition, 3) decriminalisation and 4) free market (what we have for alcohol now).

Our conclusions were published in a paper in 2018.[36]

The work was challenging; it turned out there were 27 policy outcomes we needed to consider (significantly more than the 16 harms!). When we added up the scores, it turned out that state control was the most beneficial overall, for both alcohol and cannabis. The Swedes had got it right.

THE POLICIES THAT WILL STOP US DRINKING TOO MUCH

1. Make alcohol a national health priority.
2. Tax according to alcohol content since alcohol is the dangerous drug in drinks. Everyone accepts the rationality of this between alcohol classes – e.g. sherry is taxed more than beer and less than spirits, so there is a precedent that could easily be brought into action. A can of 8 per cent lager should cost twice that of a 4 per cent one and four times that of a 2 per cent one.
3. Increase alcohol tax to bring the cost of alcohol in real terms back to where it was in the 1950s before the progressive rise in consumption started, i.e. gradually – say over five to ten years – triple the price. I have estimated that the average taxpayer would save in the order of £100 per year through the reduced costs of alcohol-related harms if we increased the price as suggested. Higher-rate tax payers would save even more.
4. Stop selling strong alcohol in supermarkets; use the Swedish model where alcoholic drinks of more than 3.5 per cent can only be sold in licensed shops with more limited opening times than supermarkets. Supermarket alcohol sales are not only destroying lives but also public houses and other alcohol outlets where drinking is

conducted in a social manner, where intoxication can be monitored and young people can learn to drink socially and more sensibly.

5. Ban special discounting of alcohol in bars e.g. happy hours, all you can drink for £10 etc.

6. Stop routinely selling wine in the large 250ml glasses that have crept up in use in recent years. Go back to smaller glasses. By law all bars are required to offer wine in 125ml (small) glasses; but in practice most of us assume that 175ml is small and 250ml large. The usual strategy in bars is to ask if people want a large glass and if they say no then give a 175ml rather than a 125ml serving. For a medium-size woman, five large glasses of wine in one hour will lead to a blood alcohol level of 300mg%, which is enough to put her in a coma.

7. Repeal the 24-hour licensing law so bars close at 11 p.m.

8. Reform organisations such as Carnage UK that seemingly encourage dangerous levels of drinking as entertainment.

9. Make it a law that all alcohol outlets must sell non-alcoholic drinks so that those who like the taste of ale, wine or spirits can enjoy them without the risk of intoxication. Sell these drinks below the cost of equivalent alcohol-containing ones and make it obvious that they are available.

10. Enforce the law that makes serving drunk customers illegal in bars: have breathalysers in bars and clubs so that seemingly intoxicated people can be tested and denied more alcohol if they are above 150mg%.

11. Add notices to all drinks warning of the damage alcohol does, as with those on cigarette packets.

12. Reduce the drink-driving limit to 50mg% to deter drink-driving and also to reduce drinking. And if people are caught, get them properly assessed for possible alcohol dependence, and repeal their licences if they flout DVLA guidance. Encourage the wider use of alcohol detectors in cars.

13. Invigorate the treatment of alcohol dependence by making alcohol a priority for the National Treatment Agency for Substance Abuse; encourage the use of proven treatments that reduce drinking and stop relapse.

14. Provide incentives to the pharmaceutical industry to develop new treatments for alcohol dependence and its consequences.

15. Encourage research into developing alcohol alternatives that are less dangerous, intoxicating and addictive than ethanol.

16. Educate children from primary school age about the dangers of alcohol. Educate their parents, too. Support children whose parents' drinking is harming them.

17. Develop public campaigns to make alcohol unfashionable, just as was done for tobacco.

18. Ban all alcohol advertising, as was done for tobacco.

19. Ban all government-supported organisations e.g. universities from having subsidised bars. Ban drinking games and pub-crawls in public organisations such as university sports and social clubs; remove financial support from clubs that allow these.

20. Finally, a measure that could be a powerful tool in the implementation of the above would be to reduce the use of alcohol by politicians. It may distort their objectivity in

law-making in relation to the harms of alcohol. Only allow them to vote if they have passed a breathalyser test. Get them to openly declare any association with the alcohol industry. The government's wine cellar should be closed and the subsidy of alcohol in the Houses of Parliament stopped. Somehow though, it seems unlikely that MPs would call time on that particular perk . . .

IS THE FUTURE SOBER?

Despite the best efforts of the drinks advertising industry, being sober is finally cool. Sober raves, sober bars, sober socialising: there's been a rise in the number of people getting together and having a good time that's not about getting drunk. University College London research analysed survey figures from 2015, and concluded that the proportion of 16- to 24-year-olds who say they never drink went up from 18 per cent in 2005 to 29 per cent in 2015.[37]

The same trend has been noted in other European countries as well as North America and Australia. Some of this is driven by people who want to get sober because they are being damaged by alcohol and the existing ways to get sober – perhaps AA or what's offered by the NHS – are not working for them.

There are also the teens and twentysomethings for whom drinking is just Not Cool. They might have seen their parents staggering around drunk and thought, yuk, that's not for me. You could see it as the new rebellion: sober as the millennial equivalent of punk, maybe. Our society has gone so far in its normalisation of extreme hedonism, of Saturday-night town centres filled with staggering, puking youths and of middle-aged people getting tanked up on any occasion from a new baby to New Year, that the highest form of rebellion a teenager can reach is to go straight. To be Saffy to our collective Edina, as in the BBC TV series *Absolutely Fabulous*.

The new sober has an underlying theme of self-help and wellness. This group are into the natural high you can get from feeling super-fit and mentally well, from realising that if you want to optimise your mental and physical health, there is perhaps no place for alcohol. Although it's also possible that some of the drop in alcohol consumption in the young has been prompted by an increased use of cannabis and other psychoactive drugs.

Social media has helped gather like-minded sober people. It's a place where people can tell compelling stories of why sober makes their life so much better. And a place where being sober is now being sold as aspirational.

As a result, big business is realising the profit in non-alcoholic and 'health-promoting' drinks and is launching new ones at an unprecedented rate.

If the trend for not drinking continues, perhaps alcohol will become a thing of the past in the next 100 years, fade into social unacceptability? Perhaps the whole conversation we are having will go away? Somehow I doubt this, because fashions are cyclical. And also because alcohol is such a powerful enabler of sociability and good humour; vital needs for most humans. I suspect the only way to reduce its use is to develop safer alternatives with similar pro-social effects.

THE ISSUE OF INDUSTRY FUNDING

Public Health England wanted to do an advertising campaign promoting (at least) two drink-free days a week.[38] This is sensible, safe and appropriate advice. But PHE doesn't have much money for public advertising so they agreed to partner with Drinkaware, a charity that's funded by the alcohol industry.

Hundreds of academics signed a petition censoring PHE. And their senior alcohol advisor Ian Gilmore resigned in protest.[39]

It wasn't the content of the campaign that was contentious, it was where the money was coming from. As a pragmatist, I don't have a problem with industry funding independent research and education, providing they don't get to have any influence on the content and outputs. But I can see why people do object. What is a shame is that the campaign didn't happen because there was no other source of funding. In reality, people will die as a result of not knowing this useful information.

IS THE FUTURE SYNTHETIC ALCOHOL?

One of the huge drawbacks of alcohol is that there is no antidote to an overdose. When you drink too much, unless you vomit, you can die of alcohol poisoning.

In 1980, when working on my research degree at Oxford University, I discovered an antidote to alcohol intoxication – at least in rats. Before, we'd known that amphetamines could to some extent sober people up, but mine was the first proof of a receptor selective antidote. Actually blocking the GABA-receptor blocked the effects of alcohol. So if you took it, it would stop you being drunk.

One problem with it was, it didn't stop the toxicity of alcohol, only the intoxicating effects. So it might have even encouraged people to drink more. And if you took too much, you would get very anxious and possibly have fits. Needless to say, it was never put into manufacture.

But what it did do was make me think about the question: what if we could design an alcohol that's not toxic?

In the past 15 years, I've been working on this with a small team. The concept is called Alcarelle and involves developing compounds that are more selective than alcohol. Our goal is to replicate the social and relaxation effects of alcohol but have less of the unwanted effects. Of course, before any new food ingredient can be sold rigorous testing is required through standard food regulations to ensure safety.[40]

I'd imagine that Big Alcohol will have very mixed feelings about the launch of a non-ethanol alternative to alcohol. As a competitor to ethanol it would require a change of strategy from them, something they haven't seemed willing to do when it comes to minimum unit pricing. I'd imagine they'd also be concerned that one of their competitors might go ahead with it and they'd be left behind. That is why we are determined to do everything we can to work in partnership with the innovators in the drinks industry, to make a new generation of better drink products. Alcarelle will focus on the science and the drinks companies will be able to license the technology, formulate their own flavouring and branding and market their own range of new-generation beverages.

We are finding alcohol companies are more interested in our

progress today and are more open to collaboration than before. This is due to a number of developing trends. The first is simply the massive move towards sustainability. The carbon footprint of alcohol is pretty large as it takes a lot of energy to shift bottles around, while Alcarelle could be shipped in a very concentrated form.

Another reason is because the young are showing signs they'd like to turn away from alcohol. The growth area of the drinks market is currently in the non-alcoholic and healthy functional drinks category and there is a real gap, with very little on offer that is truly satisfying to adult consumers. Can you imagine a health drink crossed with the relaxed and sociable benefits of alcohol – but with no loss of control or hangover?

The next step is to raise investment in order to take the molecule through extensive safety testing. Alcarelle naturally must meet the safety criteria of food standards. What seems rather irrational about this is that alcohol is absolutely proven to be extremely toxic, and would fail safety standards on every count. In fact, as I explained in Chapter Two, if alcohol went through food standards testing for toxicity now, you'd only be allowed to have less than a wine glass of alcohol a year.[41]

I have a personal interest outside of Alcarelle too, in the possible use of compounds found in nature to formulate an all-natural drink that combines efficacy on the GABA system with the same good effects as alcohol: sociability and relaxation. The advantage of this is the possibility of using known plant extracts that have been around for centuries, if not millennia, and so a much faster route to market. If successful, this would potentially provide the funding to pursue the much more ambitious synthetic solution. Finally, in theory we could bring both together: a concoction of plants and synthetics that gives the perfect buzz.

THE RUSSIAN STORY

Russia is the most phenomenal illustration of how the government controlling the amount people drink, via movement measures and price, can save thousands of lives.

Historically, Russia has had one of the highest drinking levels in the world (it's still extremely high). Russians drink so much that they die of acute alcohol poisoning before they live long enough to get liver cirrhosis!

One illustration of the enormous health damage from alcohol comes from research that showed almost half of all deaths in working-age men in a typical Russian city may be accounted for by hazardous drinking.[42]

Various governments have attempted to introduce measures to reduce the amount the population drinks, mostly because it kills so many people. And they have been able to do this because all vodka production is state-controlled.

Each time production was cut, you could see a rise in life expectancy exactly paralleled by an opposite fall in deaths from alcohol poisoning.[43]

But the Russian people didn't like this, so every time it happened, they protested. The Russian relationship with alcohol brings to mind the phrase 'opium of the people'. The policy has been instigated and reversed several times with the same results.

There's another way to consider the effect of price on consumption, too. In developing countries, as soon as people get any surplus income, they tend to spend it (or some of it) on alcohol. In other words, consumption and GDP are linked. This graph shows how alcohol consumption goes up once GDP reaches a certain level just above subsistence.

Fig 11. The link between alcohol consumption and GDP

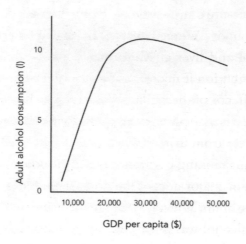

Source:
https://www.researchgate.net/publication/225040876_Global_and_Country_Specific_A
dult_per_capita_Consumption_of_Alcohol_2008

This is what is happening right now in Nigeria and South Africa and Kenya, which all have high levels of alcohol intake, as shown by a recent *The Lancet* paper.[44] It appears that if you can afford alcohol, you drink it. And, what won't surprise you is that some of the people who drink it will become alcoholics.

CONCLUSION

OVER THE PAST 50 years – my scientific lifetime – there have been massive advances in the science of alcohol. We now have a much clearer picture of how its toxic properties affect almost all areas of the body. Whereas before we just knew the link between alcohol and liver cirrhosis, we now also know the particular contribution it makes to high blood pressure, with its knock-on effects of heart attacks and strokes. And we know alcohol use is associated with at least eight cancers too. The risk of cancer from drinking is not as high as the risk of lung cancer from smoking but it is still significant.

Another area of major advance in knowledge is the actions of alcohol on the brain. When I started in medicine in the 1970s it was believed alcohol was a solvent, that it dissolved into the cell walls of neurons and so disrupted their activity. We now know this is a minor action and that what's really important is how alcohol acts on many different neurotransmitter systems. For example, alcohol enhances the activity of the sedating inhibitory neurotransmitter GABA but also blocks the activity of the wakefulness-promoting neurotransmitter glutamate. This double whammy is why severe alcohol intoxication can be so dangerous and why it can lead to coma and even death.

As well as affecting GABA and glutamate, alcohol produces a cocktail of effects on other neurotransmitters. It increases the stimulant neurotransmitter dopamine, the pleasure peptide endorphin and also the empathy-producing neuro-transmitter serotonin. It's this special cocktail that makes alcohol so pleasurable, especially socially. But it's also its effect on these neurotransmitters that can lead to dependence and addiction, especially in people with genetic predisposi-tions or those who are using alcohol to deal with stress or other mental health issues.

This book also shares with you the challenges that govern-ments and policy-makers face in dealing with the harms alco-hol produces combined with our desire to drink and the influ-ence of the immensely rich and powerful alcohol industry. I believe there are proven policies that rational governments could instigate to maximise the benefits and minimise the harms of alcohol in our society. I have campaigned all my professional life for these. The challenge is getting the media and you – the general public – to understand this and campaign for them too. I hope that this book gives you the information and motivation to support me in this ambition.

UK CHIEF MEDICAL OFFICER'S ALCOHOL GUIDELINES

YOU ARE SAFEST not to drink regularly more than 14 units per week, to keep health risks from drinking alcohol to a low level.

- If you do drink as much as 14 units per week, it is best to spread this evenly over 3 days or more. If you have one or two heavy drinking sessions (i.e. drink more than five units a session), you increase your risks of death from long-term illnesses and from accidents and injuries.
- The risk of developing a range of illnesses (including, for example, cancers of the mouth, throat and breast) increases with any amount you drink on a regular basis.
- If you wish to cut down the amount you're drinking, a good way to help achieve this is to have several drink-free days each week.

RESOURCES

Addiction Helpline www.addiction.org.uk

Adfam (families, drugs and alcohol) www.adfam.org.uk

Al-Anon www.al-anonuk.org.uk

Bottled Up www.bottled-up.com

Campaign Against Living Miserably (CALM) – for men
Call 0800 58 58 58 – 5 p.m. to midnight every day or
visit www.thecalmzone. net and click on the webchat page.

Childline – for children and young people under 19
Call 0800 1111 – the number won't show up on your
phone bill. www.childline.org.uk

Drugfam (Drug and Alcohol Addiction: families need
support too) www.drugfam.co.uk

Nacoa (National Association for Children of Alcoholics)
www.nacoa.org.uk and helpline 0800 358 3456

Papyrus (Prevention of Young Suicide) – for people under 35
papyrus-uk.org
Call 0800 068 41 41 – Monday to Friday 10 a.m. to 10 p.m.,
weekends 2 p.m. to 10 p.m., bank holidays 2 p.m. to 5 p.m.
Text 07786 209697 or email pat@papyrus-uk.org

Samaritans
Call 116 123 or email jo@samaritans.org

Scottish Families Affected by Alcohol & Drugs
www.sfad.org.uk

NOTES

INTRODUCTION

1 www.bmj.com/content/362/bmj.k3944
2 'Like tobacco, alcohol kills some users slowly' www.bmj.com/bmj/
   section-pdf/986178?path=/bmj/363/8172/This_Week.full.pdf
3 www.gov.uk/government/news/public-health-england-and-drinka-
   ware-launch-drinkfree-days
4 www.theguardian.com/society/2019/may/15/britons-get-drunk-
   more-often-than-35-other-nations-survey-finds
5 ons.gov.uk/peoplepopulationandcommunity/healthandsocialcare/
   causesofdeath/bulletins/alcoholrelateddeathsintheunitedkingdom/
   registeredin2017
6 who.int/news-room/detail/21-09-2018-harmful-use-of-alcohol-kills
   -more-than-3-million-people-each-year-most-of-them-men
7 www.alcoholchange.org.uk/alcohol-facts/fact-sheets/alcohol
   -statistics
8 digital.nhs.uk/data-and-information/publications/statistical/
   statistics-on-alcohol/2019
9 en.wikipedia.org/wiki/Weihenstephan_Abbey
10 www.bmj.com/content/362/bmj.k3944
11 www.thelancet.com/journals/lancet/article/PIIS0140-
   6736(10)61462-6/fulltext
   www.ncbi.nlm.nih.gov/pubmed/25922421
   journals.sagepub.com/doi/abs/10.1177/0269881119841569?journal-
   Code=jopa
12 doi.org/10.1002/ijc.27553

HOW DRINKING AFFECTS YOUR BODY AND BRAIN

1 www.ncbi.nlm.nih.gov/pubmed/28216062
2 Additional information from: www.alcohol.org/effects/blood-alcohol-concentration
3 pdfs.semanticscholar.org/9bb9/846bf3eec9fd-7b1a37ae6e0e8cc025d34563.pdf
4 foodanddrink.scotsman.com/drink/the-controversial-story-of-buckfasts-rise-to-prominence-in-scotland
5 www.dailymail.co.uk/news/article-3242494/Revenge-PM-s-snub-billionaire-funded-Tories-years-sparked-explosive-political-book-decade.html
6 www.ncbi.nlm.nih.gov/pmc/articles/PMC2426682
7 www.ncbi.nlm.nih.gov/pubmed/27380261
8 www.ncbi.nlm.nih.gov/pmc/articles/PMC543875
9 theconversation.com/lining-your-stomach-with-milk-before-a-big-night-out-and-other-alcohol-myths-88116
10 pubs.niaaa.nih.gov/publications/aa72/aa72.htm
11 www.researchgate.net/publication/263287740_Editorial_Can_Hangover_Immunity_be_Really_Claimed
12 www.researchgate.net/publication/26693901_Are_Some_Drinkers_Resistant_to_Hangover_A_Literature_Review
13 www.cdc.gov/media/releases/2015/p1015-excessive-alcohol.html
14 www.medicalnewstoday.com/articles/325619.php
15 www.ncbi.nlm.nih.gov/pmc/articles/PMC4230485
16 www.ncbi.nlm.nih.gov/pubmed/22434663
17 www.sciencedaily.com/releases/2018/05/180514095530.htm
18 news.bbc.co.uk/1/hi/health/8416431.stm
19 www.ncbi.nlm.nih.gov/pmc/articles/PMC5515685
20 journals.sagepub.com/doi/pdf/10.1177/2050324517741038

THE HEALTH HARMS OF ALCOHOL

1 www.researchgate.net/publication/235971838_Alcohol_consumption_alcohol_dependence_and_attributable_burden_of_disease_in_Europe_Potential_gains_from_effective_interventions_for_alcohol_dependence
2 www.sciencedaily.com/releases/2018/10/181003102732.htm
3 understandinguncertainty.org/files/2012bmj-microlives.pdf
4 assets.publishing.service.gov.uk/government/uploads/system/uploads/attachment_data/file/489795/summary.pdf
5 www.thelancet.com/journals/lancet/article/PIIS0140-6736(18)32744-2/fulltext
6 www.ncbi.nlm.nih.gov/pubmed/27073140
7 www.ncbi.nlm.nih.gov/pubmed/15770105
8 www.dailymail.co.uk/news/article-5843059/NHS-pharmacy-worker-37-died-toxic-poisoning-misjudging-alcohol-measures-holiday.html
9 www.nhs.uk/conditions/alcohol-related-liver-disease-arld/treatment
10 Personal communication from Professor Nick Sheron
11 eprints.soton.ac.uk/427505/1/1_s2.0_S246826671830241X_main.pdf
12 www.ncbi.nlm.nih.gov/pmc/articles/PMC3992057
13 www.drinkaware.co.uk/alcohol-facts/health-effects-of-alcohol/diseases/alcohol-and-liver-cancer
14 www.wcrf.org/dietandcancer/exposures/alcoholic-drinks
15 www.aicr.org/continuous-update-project/reports/breast-cancer-report-2017.pdf
16 www.wcrf.org/dietandcancer/exposures/alcoholic-drinks
17 cordis.europa.eu/project/rcn/92876/reporting/en
18 www.medicalnewstoday.com/articles/314539.php
19 www.thelancet.com/journals/lancet/article/PIIS0140-6736 18)31772-0/fulltext
20 www.vox.com/science-and-health/2018/8/29/17790118/alcohol-lancet-health-study
21 Milwood et al., 2019 Lancet on Kadoorie study
22 www.ahajournals.org/doi/full/10.1161/01.HYP.33.2.653
23 www.ahajournals.org/doi/pdf/10.1161/strokeaha.108.520817
24 www.thelancet.com/journals/lanpub/article/PIIS2468-2667(17)30003-8/fulltext
25 www.bmj.com/content/357/bmj.j2353

26 www.bmj.com/company/newsroom/both-long-term-abstinence-and-heavy-drinking-may-increase-dementia-risk
27 www.ahajournals.org/doi/10.1161/CIRCULATIONAHA.108.787440
28 www.nytimes.com/2018/06/18/health/nih-alcohol-study.html
29 www.ncbi.nlm.nih.gov/pubmed/26294775
30 www.ncbi.nlm.nih.gov/pubmed/29103170
31 www.ncbi.nlm.nih.gov/pmc/articles/PMC4590619

AAA: ALCOHOL, ACCIDENTS AND AGGRESSION

1 www.theguardian.com/society/2019/sep/04/violence-nhs-staff-face-routine-assault-intimidation
2 assets.publishing.service.gov.uk/government/uploads/system/uploads/attachment_data/file/827834/drink-drive-final-estimates-2017.pdf
3 www.gov.iruk/government/statistics/reported-road-casualties-in-great-britain-final-estimates-involving-illegal-alcohol-levels-2017
4 webarchive.nationalarchives.gov.uk/20100921035247/http:/northreview.independent.gov.uk/docs/NorthReview-Report.pdf
5 www.ncbi.nlm.nih.gov/pmc/articles/PMC4814299
6 www.alcoholhelpcenter.net/Program/BAC_Standalone.aspx
7 www.researchgate.net/publication/223136111_Alcohol-Related_Risk_of_Driver_Fatalities_An_Update_Using_2007_Data
8 ibid
9 webarchive.nationalarchives.gov.uk/20100921035247/http:/northreview.independent.gov.uk/docs/NorthReview-Report.pdf
10 www.sussex.ac.uk/broadcast/read/48912
11 etsc.eu/wp-content/uploads/report_reducing_drink_driving_final.pdf
12 www.ias.org.uk/uploads/pdf/bloodalcoholcontenteffectivenessreview.pdf
13 eprints.gla.ac.uk/189646
14 www.oisevi.org/a/archivos/estudios-especificos/ong/Union-Europea-Druid-Final-Report.pdf
15 www.ncbi.nlm.nih.gov/pubmed/18955613

16 www.bath.ac.uk/announcements/why-the-effects-of-a-boozy-bank
    -holiday-binge-could-last-longer-than-you-think
17 www.ncbi.nlm.nih.gov/pubmed/20497803
18 en.wikipedia.org/wiki/The_Gulag_Archipelago
19 www.historyextra.com/period/viking/the-truth-about-viking
    -berserkers
20 www.sciencedaily.com/releases/2009/02/090218081624.htm
21 www.inverse.com/article/49228-who-alcohol-report-drinking
    -deaths
22 www.drinkaware.co.uk/research/data/consequences
23 www.sciencedirect.com/science/article/pii/S1353113105000520
24 www.bbc.co.uk/news/uk-england-20630771
25 metro.co.uk/2011/06/17/ascot-fight-on-ladies-day-stuns-horse-
    racing-fans-47597/
26 alcoholchange.org.uk/publication/are-you-looking-at-me
27 theconversation.com/if-england-gets-beaten-so-will-she-the-link-
    between-world-cup-and-violence-explained-99769
28 www.ias.org.uk/uploads/IAS%20report%20Alcohol%20domes-
    tic%20abuse%20and%20sexual%20assault.pdf

WHAT ALCOHOL DOES TO YOUR MENTAL HEALTH

1 www.nhs.uk/news/mental-health/alcohol-and-depression
2 www.cmaj.ca/content/191/27/E753
3 pubs.niaaa.nih.gov/publications/arh26-2/130-135.htm
4 adaa.org/understanding-anxiety/social-anxiety-disorder/social-
   anxiety-and-alcohol-abuse
5 www.cambridge.org/core/journals/the-british-journal-of-psychia-
   try/article/loss-of-consciousness-and-posttraumatic-stress-disor-
   der/E82CD29771CBD412EBA3D51A896CB059
6 www.ncbi.nlm.nih.gov/pmc/articles/PMC3770804
7 www.ncbi.nlm.nih.gov/pubmed/29183241
8 academic.oup.com/qjmed/article/99/1/57/1523792
9 www.verywellmind.com/adult-alcoholism-adhd-connected-63078
10 www.sciencedirect.com/science/article/abs/pii/088761859290009V
11 jamanetwork.com/journals/jamapsychiatry/article-abstract/497548

DRINK?

12 www.sciencedirect.com/science/article/abs/pii/S0306460398000094
13 www.ncbi.nlm.nih.gov/pubmed/6598815
14 www.sciencedirect.com/science/article/abs/pii/S0306460396000536
15 link.springer.com/article/10.1007/s00127-005-0981-3
16 www.ncbi.nlm.nih.gov/pubmed/22983943

HORMONES AND FERTILITY

1 www.ncbi.nlm.nih.gov/pmc/articles/PMC1772851
2 www.ncbi.nlm.nih.gov/pmc/articles/PMC3527168
3 www.sciencemediacentre.org/expert-reaction-to-research-on-alcohol-intake-and-male-fertility
4 www.bionews.org.uk/page_137273
5 www.sciencedaily.com/releases/2014/10/141002221232.htm
6 ibid
7 www.bmj.com/content/354/bmj.I4262
8 www.nature.com/articles/s41598-017-14261-8
9 www.ncbi.nlm.nih.gov/pubmed/3367299
10 www.sciencedaily.com/releases/2017/04/170411090205.htm
11 www.nhs.uk/conditions/pregnancy-and-baby/alcohol-medicines-drugs-pregnant
12 pubs.niaaa.nih.gov › publications › arh22-1
13 www.sciencedaily.com/releases/2018/11/181128154000.htm
14 www.nhs.uk/conditions/foetal-alcohol-syndrome
15 bmjopen.bmj.com/content/8/12/e022578
16 www.ncbi.nlm.nih.gov/pmc/articles/PMC5609722
17 www.ncbi.nlm.nih.gov/pmc/articles/PMC4445685
18 en.wikipedia.org/wiki/The_South_African_Wine_Initiative#cite_note-AFP-2
19 www.theguardian.com/global-development/2018/may/27/mothers-children-foetal-alcohol-syndrome-south-africa
20 www.sciencedaily.com/releases/2017/08/170821122756.htm
21 ibid
22 www.cochrane.org/CD011445/PREG_ethanol-alcohol-preventing-preterm-birth
23 academic.oup.com/humupd/article/22/4/516/2573866

NOTES

24 www.ncbi.nlm.nih.gov/pubmed/16500807
25 www.breastcancer.org/research-news/20080311
26 www.sciencedaily.com/releases/2017/11/171107092906.htm
27 www.ncbi.nlm.nih.gov/pubmed/24118767
28 www.sciencedirect.com/science/article/abs/pii/S0306453011003350
29 www.pnas.org/content/pnas/early/2009/06/26/0812809106.full.pdf
30 www.ncbi.nlm.nih.gov/pubmed/30738971

HOW ALCOHOL AFFECTS YOUR QUALITY OF LIFE

1 www.nhs.uk/apps-library/sleepio
2 science.howstuffworks.com/life/sleep-obesity1.htm
3 www.sciencedaily.com/releases/2017/04/170420114020.htm
4 jech.bmj.com/content/71/12/1177
5 www.sussex.ac.uk/broadcast/read/47131
6 www.independent.co.uk/life-style/health-and-families/health-news
/the-incredible-shrinking-man-5544531.html
7 www.euro.who.int/-data/assets/pdf_file/0018/319122/Public-
health-successes-and-missed-opportunities-alcohol-mortality-
19902014.pdf
8 www.researchgate.net/publication/26314858_Beer_consumption_
and_the_'beer_belly'_Scientific_basis_or_common_belief
9 www.thelancet.com/pdfs/journals/lanpub/PIIS2468-2667(17)30089
-0.pdf
10 https://www.alcoholpolicy.net/2019/06/a-new-report-from-the-
institute-of-alcohol-studies-ias-says-the-cost-of-hangovers-is-up-to
-14-billion-a-year-with-as-ma.html
11 www.sciencedaily.com/releases/2018/08/180825124245.htm
12 medicalxpress.com/news/2019-08-hangovers-brain-function.html
13 www.drinkaware.co.uk/alcohol-facts/health-effects-of-alcohol/
lifestyle/can-alcohol-affect-sports-performance-and-fitness-levels
14 www.ncbi.nlm.nih.gov/pmc/articles/PMC4629692
15 www.telegraph.co.uk/men/thinking-man/tony-adams-life-alco-
holic-knew-play-football-didnt-know
16 www.bbc.co.uk/news/uk-england-manchester-47864761

17 www.independent.co.uk/sport/football/premier-league/arsenal-news-nitrous-oxide-laughing-gas-latest-ozil-lacazette-guendouzi-aubameyang-video-club-a8672061.html

18 www.dailymail.co.uk/sport/football/article-3037536/Raheem-Sterling-filmed-taking-hippy-crack-just-days-pictures-Liverpool-star-smoking-shisha-pipe.html

19 www.theguardian.com/science/sifting-the-evidence/2015/nov/17/sex-does-alcohol-really-make-you-better-in-bed

20 www.ncbi.nlm.nih.gov/pmc/articles/PMC1403295

21 www.sciencedaily.com/releases/2016/08/160804141034.htm

22 www.sciencedaily.com/releases/2016/09/160918214439.htm

23 www.sciencedirect.com/science/article/pii/S205011611830120X

24 www.ncbi.nlm.nih.gov/pmc/articles/PMC2917074

25 www.sciencedaily.com/releases/2014/03/140328102907.htm

26 www.forbes.com/sites/chunkamui/2016/03/22/wine-and-sleep-make-for-better-decisions/#38f815a824b1

27 www.sciencedirect.com/science/article/pii/S1053810016303713

28 www.forbes.com/sites/chunkamui/2016/03/22/wine-and-sleep-make-for-better-decisions/#38f815a824b1

29 en.wikipedia.org/wiki/Rum_ration

30 www.telegraph.co.uk/sport/othersports/darts/6637884/Andy-Fordham-I-became-world-darts-champion-despite-never-being-sober.html

31 www.dailymail.co.uk/news/article-2794310/drunk-belgian-anaesthetist-killed-british-mother-caesarean-told-police-need-vodka-don-t-shake.html

32 www.theguardian.com/world/2014/oct/03/british-woman-dies-botched-caesarean-france

33 www.bmj.com/content/2/5103/993

34 DOI: 10.1177/0269881117735687

ADDICTION: HAVE I GOT AN ALCOHOL PROBLEM?

1 pdfs.semanticscholar.org/60cd/d49b5ed6a84423762a366727ae4880c6f255.pdf

2 www.drinkaware.co.uk/alcohol-facts/health-effects-of-alcohol/alcohol-and-gender/alcohol-and-men

NOTES

3 psycnet.apa.org/record/1977-27694-001
4 americanaddictioncenters.org/alcoholism-treatment/symptoms-and-signs/hereditary-or-genetic
5 www.ncbi.nlm.nih.gov/pubmed/28734091
6 www.ncbi.nlm.nih.gov/pubmed/8611056
7 ibid
8 www.ncbi.nlm.nih.gov/pubmed/15704214
9 www.alcohol.org/alcoholism/is-it-inherited
10 en.wikipedia.org/wiki/Disulfiram-like_drug
11 americanaddictioncenters.org/alcoholism-treatment/symptoms-and-signs/hereditary-or-genetic
12 www.imperial.ac.uk/news/192468/scientists-untangle-links-between-genes-intake
13 www.jsad.com/doi/pdf/10.15288/jsa.1977.38.1004
14 www.cochrane.org/CD004148/ADDICTN_effectiveness-brief-alcohol-interventions-primary-care-populations
15 www.ncbi.nlm.nih.gov/pubmed/11912067
16 www.niaaa.nih.gov/news-events/news-releases/niaaa-reports-project-match-main-findings
17 ajp.psychiatryonline.org/doi/full/10.1176/appi.ajp.2016.16050589
18 www.nus.org.uk/en/news/press-releases/new-survey-shows-trends-in-student-drinking
19 www.independent.co.uk/news/uk/home-news/student-death-initiation-ceremony-newcastle-university-ed-farmer-drunk-a8602061.html
20 www.ncbi.nlm.nih.gov/pmc/articles/PMC3286419

THE SOCIAL BENEFITS OF ALCOHOL

1 www.sirc.org/publik/drinking3.html
2 medovina.com/history.htm
3 Roueche, B. 'Alcohol in Human Culture'. In: Lucia, S., (ed.) *Alcohol and Civilization*. NY: McGraw-Hill, 1963, p.171.
4 www.ncbi.nlm.nih.gov/pubmed/21954999
5 medium.com/pebble-research/the-secret-to-happiness-2348011c8497#.q95rbpl1m
6 link.springer.com/article/10.1007/s40750-016-0058-4

DRINK?

7 eprints.qut.edu.au/83729/4/Ryan_McAndrew_Thesis.pdf
8 www.ncbi.nlm.nih.gov/pubmed/3981386
9 www.sciencedaily.com/releases/2017/05/170515091131.htm
10 journals.sagepub.com/doi/10.1177/1741659005057641
11 www.psychologicalscience.org/news/releases/moderate-doses-of-alcohol-increase-social-bonding-in-groups.html
12 news.bbc.co.uk/1/hi/health/3035442.stm
13 www.kent.ac.uk/news/society/9730/alcohol-can-make-you-momentarily-happier
14 www.sciencedirect.com/science/article/abs/pii/S0277953616301344
15 www.sciencedaily.com/releases/2017/05/170515091131.htm
16 www.sciencedaily.com/releases/2011/12/111219135215.htm
17 www.sciencedaily.com/releases/2018/07/180705114017.htm
18 www.ncbi.nlm.nih.gov/pubmed/30717614
19 www.sciencedaily.com/releases/2018/07/180705114017.htm

HOW TO DRINK THE WAY YOU WANT TO (AND SENSIBLY)

1 drinktank.org.au/2018/03/there-are-four-types-of-drinker-which-one-are-you
2 graziadaily.co.uk/life/food-and-drink/heres-reason-sometimes-get-drunk-quickly
3 www.sciencedaily.com/releases/2008/07/080718180723.htm
4 www.researchgate.net/publication/328678730_Young_Adults_Do_Not_Catch_Up_Missed_Drinks_When_Starting_Later_at_Night-An_Ecological_Momentary_Assessment_Study
5 www.ncbi.nlm.nih.gov/pmc/articles/PMC5344720
6 www.ncbi.nlm.nih.gov/pubmed/27419377
7 www.ncbi.nlm.nih.gov/pubmed/17344205 and also www.publications.parliament.uk/pa/cm201012/cmselect/cmsctech/1536/153602.htm
8 www.sussex.ac.uk/broadcast/read/47131

CHILDREN AND ALCOHOL

1 www.mirror.co.uk/news/uk-news/drunken-student-charged-after-urinating-425067
2 digital.nhs.uk/news-and-events/latest-news/pupils-who-have-recently-smoked-drunk-alcohol-and-taken-drugs-are-more-likely-to-be-unhappy-according-to-new-report
3 John Moores University. Updating England-Specific Alcohol-Attributable Fractions report, 2013.
4 www.theguardian.com/uk-news/2018/oct/22/newcastle-student-died-after-initiation-bar-crawl-inquest-told
5 www.premierleague.com/news/1291167
6 www.dailymail.co.uk/news/article-1309143/Euan-I-young-pints-drunk-I-lost-Family-friend-night-Blairs-son-arrested.html
7 Nutt, D. J., *Drugs: Without the Hot Air*. UIT Press, 2012.

IS THERE A SOLUTION TO THE BOOZE CRISIS?

1 www.telegraph.co.uk/news/2018/08/23/alcohol-revenue-would-decline-13-billion-drinkers-stuck-recommended
2 tobaccocontrol.bmj.com/content/11/suppl_1/i110
3 en.wikipedia.org/wiki/British_Doctors_Study
4 www.independent.co.uk/news/blair-launches-war-on-drug-abuse-1262363.html
5 www.ave.ee/download/Alcohol%20England.pdf
6 www.ias.org.uk/What-we-do/Alcohol-Alert/Issue-3-2004/Disabling-the-public-interest-Alcohol-Strategies-and-Policies-for-England.aspx
7 ibid
8 academic.oup.com/bmb/article/123/1/149/3958774
9 www.alcohol-focus-scotland.org.uk/news/half-of-alcohol-being-sold-under-50p-per-unit
10 www.ifs.org.uk/publications/13826
11 ibid
12 www.ias.org.uk/uploads/pdf/IAS%20reports/Dead%20on%20

Arrival_x3F_%20Evaluating%20the%20Public%20Health%20
Responsibility%20Deal%20for%20Alcohol.pdf

13 bmjopen.bmj.com/content/7/5/e013497

14 www.gov.uk/tax-on-shopping/alcohol-tobacco

15 www.gov.uk/government/publications/rates-and-allowance-excise-
duty-alcohol-duty/alcohol-duty-rates-from-24-march-2014

16 www.sciencedirect.com/science/article/pii/S0047272718302329?
via%3Dihub

17 www.ias.org.uk/What-we-do/Alcohol-Alert/Issue-2-2012/Alcohol-
problems-account-for-a-quarter-of-Scottish-intensive-care-unit-
admissions.aspx

18 www.alcoholpolicy.net/2019/06/encouraging-signs-as-consumption
-falls-one-year-after-scotlands-mup-introduction.html

19 www.bmj.com/content/366/bmj.l5274

20 eucam.info/regulations-on-alcohol-marketing/france

21 www.icadtsinternational.com/files/documents/2004_054.pdf

22 fasdprevention.wordpress.com/2011/05/12/fasd-prevention-in
-france

23 observer.com/2016/10/booze-at-the-palace-british-parliaments-30-
bars-for-thirsty-mps

24 www.theguardian.com/world/2007/feb/06/gender.politics

25 www.independent.co.uk/news/people/popular-mp-jim-dobbin-died-
during-european-trip-after-drinking-polish-vodka-9993581.html

26 www.mirror.co.uk/news/gallery/annies-bar-house-of-commons
-5165526

27 journals.sagepub.com/doi/full/10.1177/2050324518789219

28 www.ias.org.uk/What-we-do/Alcohol-Alert/Issue-3-2004/
Disabling-the-public-interest-Alcohol-Strategies-and-Policies-for-
England.aspx.

29 journals.sagepub.com/doi/full/10.1177/2050324518789219

30 www.theguardian.com/society/2012/jul/19/mps-tighter-rules-alco-
hol-ads-1

31 counseling.northwestern.edu/blog/decade-of-drunk-lyrics-how-
often-pop-music-mentions-alcohol

32 www.ias.org.uk/uploads/pdf/News%20stories/alicerap-policy-
paper.pdf

33 www.concealedwines.com/business-opportunities-scandinavia-
wine-producers/domestic-alcohol-policy-sweden-systembolaget

34 www.who.int/substance_abuse/publications/global_alcohol_
report/profiles/swe.pdf

NOTES

35 www.who.int/substance_abuse/publications/global_alcohol_
report/profiles/gbr.pdf

36 www.ncbi.nlm.nih.gov/pubmed/29459211

37 www.nhs.uk/news/lifestyle-and-exercise/young-people-turning-
their-backs-alcohol

38 publichealthmatters.blog.gov.uk/2018/09/10/why-public-health-
england-is-working-with-drinkaware-to-reduce-the-harms-of
-alcohol

39 www.alcoholpolicy.net/2018/09/phe-drinkaware-policy-reaction-
debate.html

40 alcarelle.com

41 cordis.europa.eu/project/rcn/92876/reporting/en

42 www.ncbi.nlm.nih.gov/pubmed/17574092

43 www.sscnet.ucla.edu/polisci/faculty/treis-man/Papers/Death%20
and%20Prices%20Final%20 Sept%2009.pdf

44 www.thelancet.com/journals/lancet/article/PIIS0140-
6736(18)32744-2/fulltext

INDEX

INDEX

INDEX

ACKNOWLEDGEMENTS

Most important thanks are to Brigid Moss for her great research and writing skills. Also, to Rachel Mills for finally working out what was the best book for me to extract from my research and clinical career.

Also to my many friends and colleagues who have taught me about alcohol over many decades but especially, Christer Allgulander, Ilana Crome, Dora Duka, Ian Gilmore, David Goldman, Tony Goldstone, Paul Glue, George Koob, Anne Lingford-Hughes, Markku Linnoila, Richard Lister, Jurgen Rhem, Marcus Munafo, Charles O'Brien, Robin Room, Nick Sheron, Julia Sinclair and Wim van den Brink. And for breakthrough insights in harm assessment and communication, Larry Philips and David Spiegelhalter.

And my wife Di who has for 40 years put up with my chaotic academic and clinical life; and our daughter Suzy, who runs Abbotshill, our wine bar in Ealing, and has taught me the value of quality organic wine!

ABOUT THE AUTHOR

DAVID NUTT
MBBChir (Cambridge) DM (Oxford), FRCP, FRCPsych,
FMedSci, hon DLaws (Bath)

David Nutt is a psychiatrist and the Edmund J. Safra Professor
of Neuropsychopharmacology in the Division of Brain
Science, Dept of Medicine, Hammersmith Hospital, Imperial
College London. His research area is psychopharmacology –
the study of the effects of drugs on the brain, from the
perspectives of both how drug treatments in psychiatry and
neurology work, and why people use and become addicted to
some drugs, such as alcohol. To study the effects of drugs in
the brain he uses state of the art techniques, such as brain
imaging with PET and fMRI plus EEG and MEG. This
research output has led to over 500 original research papers
and puts him in the top 0.1% of researchers in the world. He
has also published a similar number of reviews and book
chapters, eight government reports on drugs and 35 books,
including one for the general public, *Drugs: Without the Hot
Air*, that won the Transmission Prize in 2014.

He was previously President of the European Brain Council,
the British Association of Psychopharmacology, the British

Neuroscience Association and the European College of Neuropsychopharmacology. He is currently Founding Chair of DrugScience.org.uk a charity that researches and tells the truth about all drugs, legal and illegal, free from political or other interference. He also currently holds visiting Professorships at the Open University and University of Maastricht.

David broadcasts widely to the general public both on radio and television. In 2010, *The Times Eureka* science magazine voted him one of the 100 most important figures in British Science, and he was the only psychiatrist on the list. In 2013, he was awarded the John Maddox Prize from Nature/ Sense about Science for standing up for science and in 2017 a Doctor of Laws hon causa from the University of Bath.

en.wikipedia.org/wiki/David_Nutt
www.sciencemag.org/content/343/6170/478.full
www1.imperial.ac.uk/departmentofmedicine/divisions/ brainsciences/psychopharmacology

books to help you live a good life

Join the conversation and tell
us how you live a #goodlife

🐦 @yellowkitebooks
📘 YellowKiteBooks
📌 Yellow Kite Books
📷 YellowKiteBooks